THE WINDMILL

Sprowston Mill

THE WINDMILL

A BOOK FOR BOYS

(Young and Old)

by

W'Cdr. H. C. Harrison

A.R.C.Sc., O.B.E.

1st Edition: Melbourne, Australia, 1947

2nd Edition: Windsor, UK, 2007

Dedication of the 1947 Edition

To My Father

William Albert Harrison

One of the last of the old-time millers, whose windmill was to him a legacy of clever construction and skilled workmanship from a foregone generation ever to be regarded with esteem for its courage and accomplishment.

Notes to the 2007 Edition

This book was originally published privately in 1947 in an edition of about 100 copies. It was mainly given to the author's brothers and sisters, "the Miller's children" referred to in the text, and to their own children. The copies of the book have helped keep the story of Sprowston Mill alive down the generations since then.

Today, 60 years after it was first published, the introduction of computer scanning and digital printing techniques have made it possible to digitise the original book and re-publish it economically. It is offered for the enjoyment of a new generation for whom windmills are occasional curiosities dotted around the landscape, no longer commercially viable but in some fortunate cases restored and operated by enthusiasts.

Until the arrival of the steam engine, windmills were the most complicated machines ever built. Their construction and operations, together with asides about the families who lived with and loved their windmills, are explained beautifully here by the author Herbert Harrison, who grew up in and around Sprowston Mill on the outskirts of Norwich,

Herbert Harrison was my great-uncle and this 2007 edition is dedicated to his memory. My mother remembers him fondly and I am sure that in our current more equality-minded age he would have recognised that girls like her loved windmills too and would have included them in his subtitle! I wish that I had met him.

Simon Eccles
simon@ecclesnet.com
Windsor, 2007

Technical notes: An original 1947 book was scanned on an Epson Perfection V750 Pro scanner and the type converted to editable form by Abby FineReader 5, an optical character reading program. The halftone photographic plates were scanned in high resolution and then de-screened using Adobe Photoshop CS3. Pages were laid out using Adobe InDesign CS3 on an Apple Macintosh computer, reproducing the original type styles and layouts where possible. The layout was exported as a PDF 1.4 file.

This book was printed on-demand by a digital press. The printer and press type vary according to how the book was ordered.

2007 edition produced by Simon Eccles by kind permission of the Herbert Harrison estate.

ISBN 978-1-84799-844-6

Published by Lulu, 2007

www.lulu.com   ID: 913729

Contents

List of Illustrations

Foreword

❖

This small book is produced in the hope and expectation that it may be of interest to boys who have some leaning to mechanics and engineering. The writer is one of the numerous "miller's youngsters" who, probably unknowingly, got his first inspirations in engineering from some of the intriguing mechanisms in his father's windmill and later adopted the engineering profession as a lifetime activity.

If he has been able to describe the workings of the mill in such a manner as to be attractive to boys, this book may serve them in some sort of substitution for those things which filled so large a part of his own boyhood – a full-scale windmill to explore, to climb in and over, to help with in the running, painting and repairing – the sack hoist, a perpetual source of fun, to use for boy freight with his brothers at the frequent risk of broken limbs or neck.

The mechanisms described are those installed in the windmill depicted in Crome's picture "A Mill on Mousehold Heath", hanging in the National Gallery, London, but they are representative of all post mills. The mechanism of the tower mill is similar again, although, owing to the greater space available, the millwrights were not driven to quite such ingenious shifts to fit in all the necessary machinery and gear. In the post mill things had to be done the hard way. The tower mill, by comparison, was easy.

H. C. HARRISON,
"Mainsail Haul",
Esperance Avenue,
Middle Brighton,
Melbourne, Australia.

Domestic Prelude

As matters of domestic concern, it was originally intended that these notes should take their place at the end of this small volume. I have been prevailed upon, however, to bring them near the beginning as explanation of how the book came to be written.

The author's father, Mr. W. A. Harrison, who has now entered his tenth decade and lives at 415 London Road South, Lowestoft, Suffolk, England, inherited the Sprowston windmill from his maternal uncle, George Robertson, who himself had two miller brothers, Robert and William. It was the common custom for successive owners to engrave their name and the year of entering possession upon the "post" inside the mill. The Robertson name dated back to 1780. The author's great-grandfather, Robert Robertson, met his death in the mill in the year 1842. He became entangled in the drive from the head wheel to the sack hoist.

To my father I am indebted for much detail in the preparation of this book. The mill was to him a part of his very being. He went to Sprowston in 1884 and worked the mill continuously till 1920, when it was hired from him by the author's younger brother, Horace George Harrison, who still lives at Sprowston and was the last person to work the mill. It reverted to my father again in 1928, by which time its usefulness as a business asset had departed.

A little later it was arranged that the mill should be placed in the care of the Norfolk Archaeological Trust. Some delay occurred in regard to the access road, but presently this was settled upon, and the date for transfer was fixed as March 25th, 1933. On the previous day the wind blew strong on the mill hill, a wind such as the mill would have revelled in, had her millstones been thrown into gear and her shutters drawn. Some misguided person chose this particular day to burn off a copse of gorse brushwood on the property to the windward side of the mill hill. The stationary sails were bombarded with showers of sparks and burning brush. The canvas-covered shutters became ignited and soon the entire wooden structure. The millstones and gearing crashed into the blazing round-house. Sprowston mill would turn no more.

To my elder brother, William Edmund Harrison, I am indebted for much precision detail of the working of the mill mechanism. He now lives at Canning Vale, Warwick, Queensland. After leaving school he joined my father and became his miller until he left for Australia in 1910. He was always as much intrigued as I was with the skill and workmanship built into the mill and its gearing, and studied it closely during his period as miller.

He has told me that every joint in her was a monument to the old-time millwrights. All the main timbers were of oak. This hard and tough material had to be shaped with an adze and the joints formed by hand. It is difficult to reproduce the effect in drawings which are intended to display the working of the mechanism.

I may, perhaps, be permitted to refer to the pronounced difficulty of compiling in Australia a volume, the subject of which once existed (but no longer exists) at the opposite end of the world. With my brothers, however, I spent a great deal of time in the mill as a youngster, and the memories of those formative .years are indeed indelible, down to matters of minute detail. I have further been fortunate in being able to refresh those memories on several return visits to England since coming to Australia. It would nevertheless have been impossible to attempt the illustration of the present volume from the unfinished and inadequate sketches which I was able to draw. My illustrator, Mr. Graham R. Nisbet, has taken in hand this unfinished material and has lived himself into the windmill by piecing together every little scrap of detail from photographs and pictures. The resulting illustrations, representing much tedious work, are excellent both in their clarity and accuracy of detail.

The photographs of typical post, tower and smock mills were kindly sent to me by Mr. Rex Wailes, of Davidge Cottage, Knotty Green, Bucks – an acknowledged enthusiast and authority on windmill lore.

One other circumstance may be mentioned as having been helpful in describing, at this distance of space and time, the miller's way of life and his windmill. Of seven brothers in my father's family, four were millers – the three eldest and the youngest. The eldest, Robert William, at one time had the tower mill at Oulton Broad, Suffolk, and subsequently relinquished it to the youngest, Clifford Rockhill. The second brother, George William, owned the post mill at Gisleham, Suffolk. When, with my brothers, I made long holiday visits to these uncles, it was natural to explore their windmills and make comparison with the home mill at Sprowston.

The generation represented by the author's father and uncles may be said to be the last of the old-time millers. Bread made from millstone flour – rich, palatable and nutritious – has been superseded by bread made from the overdressed, superfine white flour of the modern roller mill. The change-over had already commenced at the period of which I write, which will explain how neither my brothers nor I adopted millering as a lifetime occupation.

Introduction

It would not be too extravagant to say that English history was influenced over a period of several centuries by the development of the windmill. We have to go back to the days of the early Plantagenets – to Richard Lion Heart – to find the first indication that it had come into existence. And then our indication takes the form of a little picture embellished in a church psalter, a curiously-displayed emblem in the stained glass of one of its windows, or again, engraved upon a brass memorial set in its wall. From that period to the beginning of the present century – nearly 600 years – the windmill determined largely where and how men should live.

But let us first go back yet another 1000 years – to the days of the Roman occupation – and see the predecessor of the windmill – the watermill – emerge. It was probably first used on the banks of a swift-running stream where large stones could be set to guide the flow of water on to its rough paddles. Roman soldiery with their chariots were then to be seen on the long straight highways which they had built to link up the larger towns, but, although an advanced and even lavish civilization was developed in the towns, life in the deep countryside persisted much as before; even the language was little corrupted. The sparse, scattered population was gathered in tiny villages along the banks of the rivers and streams, whose waters provided both drink and communication. Fresh water was not very plentiful on the uplands, and travel by foot or horse was difficult through the rude tracks of forest and fen.

The basic necessities of life were as they remain to-day – shelter, clothing and food. For shelter there were the smaller trees of the forest, birch and larch, from which the villagers built huts and thatched them. Although weaving of flax and wool was known, skins of their own sheep or the wild creatures of the woods were much valued as clothing, even as they are today. Their food was largely meat and fish from the stream, but where natural clearances occurred in the forest glades small patches of ground were tilled by means of an iron-shod wooden plough.

We may well imagine what store would be set upon the few measures of grain garnered from these tiny patches. Everyone – man, woman and child – had taken his share in their production. If the plough was guided by the strong, the oxen had to be led by the weaker. Small, shrill voices would scare the birds on summer mornings, and little bare legs would chase away the too venturesome coneys or snare them if they could. At harvest, all would be called upon to pick the ears by hand, and then would come the threshing with stick and flail. A measure of corn was hardly won and treasured accordingly against the winter months, when food was scarcer than in summer. But hardy and all as were these early folk, they could not eat the corn without some form of preparation. Doubtless it was frequently eaten in the form of a pulse made by soaking and boiling it.

However, the processes of crushing and grinding were known at that time. Crushing was the simpler and was done by an elementary form of pestle and mortar. For a mortar, a largish stone was used, having a basin-like hollow, and into this the corn was poured, a little at a time, and pounded by means of another rounded stone held in the hand. Only so much would be crushed as was soon to be eaten.

Here and there were more fortunate folk who possessed a quern for actually grinding the corn. But a quern was fashioned only by the exercise of much skill and patience. Many of them came from the workings in North Wales, where the quartzite of the mountain screes was laboriously chipped and shaped. A quern comprised a pair of round stones, a foot or more in diameter, one nesting in the other. The bottom one was formed with a shallow depression on its top face, into which fitted the rounded face of the upper. The upper stone also had a hole or eye through the centre, and another smaller blind hole near the rim into which a wooden peg was slipped for the purpose of turning it. It is evidence of the resourcefulness of these early people that they could drill a hole through hard quartzite. This was done by pressing the end of a wooden stick on to the stone and twirling it the while by means of a bow string. Wet sand was fed in to give an abrasive action and so wear away the hard stone.

Grinding the corn must have been a somewhat tedious process. Whilst the upper quern stone was turned with one hand, a little corn was dribbled into the eye with the other. After much turning the upper stone would be tilted and the coarse meal scraped out. The quern was the direct forerunner of the much-later millstones.

These early people had, then, by this time done much to alleviate their hardships. Their simple life by river vale and forest glade must have been far from unpleasant. At the same time, the business of living was always strenuous. We can almost imagine the wonderment and excitement amongst the villagers on that day when the first waterwheel became a reality.

Ethelbert, the bright, was always restless as he watched the bigger children at the quern – so much effort for so little result. And then came the day on which he took his canoe downstream to the ford by the Icknield Way to barter a sheepskin for a water pitcher. A cohort of Roman soldiery had lately passed along the Way and a charioteer had come to disaster on the boulders in midstream. Amongst the wreckage was one of the wheels of the chariot, still in reasonably sound condition, and Ethelbert had loaded it into the canoe before returning to the village. On his way back, upstream, he vaguely meditated upon the wheel rolling smoothly along the hard highway and the strong pull of the water on his own paddle. A wheel could not roll on the water, of course, but what if you lashed the paddle to the wheel? There were six spokes so you would require six paddles. He would see about it when he got back.

Many days went by, as with helping hands he built rude trestles at the water's edge to support his wheel. Then he had the further problem, of coupling it up to drive the quern. Eventually this was also solved, and now the long, tedious hours at the quern were over. In a little while word was passed up-stream and down, and presently other watermills began to make their appearance where the water in the stream ran swiftly.

Whether or no the earliest watermill was developed in exactly this way, there were now even stronger reasons than previously for people to find their living in the river valleys. Until the windmill emerged a thousand years later, there was, except for the hand quern, no other way of grinding corn.

The Norman Conquest brought many changes which affected the domestic life of the people; not all their activities were absorbed in the struggles between Kings, Earls and Bishops with which the history of the period abounds. There were castles with embattled parapets to be built and cathedrals with exquisite traceries of stone. Vellum was to be embossed with such delicate precision that a lifetime was not too long for the completion of a book. Master craftsmen were to establish their guilds tor the further advancement of their arts. But always there was the problem of living, and now food had to come to those who had taken no part in its production – the masons, the weavers, and all the skilled craftsmen of the guilds.

Under the feudal system the farm lands were held by the Barons and Lords of the Manor and were tilled by the stout-hearted villeins who owed allegiance to My Lord. How were the demands of the Sheriff to be met when he came riding in with his retinue at St. Swithins and Michaelmas to collect the King's scrutage; and what was to be traded to the guilds for their woven cloth and tanned leather jerkins unless the land was made to yield corn in abundance?

The townsfolk must have eaten much of it in the form of pulse and porridge, for watermills could only be built by swift-running water, and long distances could not be travelled over the soft rutted country lanes.

There was, then, a growing need for an increased acreage of corn-growing land and equally for more mills in which to grind the corn. Other villages, surrounded by arable fields, gradually came into existence on the uplands where water for man and beast was now provided by the sinking of wells and the excavation of ponds.

Here, if there was no running water, the wind blew stronger than it did in the valleys. There was difficulty, however, with the mills. There were some 7500 watermills recorded in the Domesday Book, but when all the suitable places on the streams had been used up no further expansion was possible. Some such influences may have guided the people on the Continent of Europe to the uplands and have urged them to develop the windmill, for it was from the Continent that the idea of the windmill was first introduced into England.

Many a Crusader made his way back to England across the Continent – as did the Lion Heart himself, getting imprisoned on the way – and would have heard of the new mill for grinding corn. We should not lose sight of the possibility, however, that a complete windmill may have been dismantled and transported over to England by ship.

If the waterwheel, in its day, called for great ingenuity on the part of its originator, the windmill called for even greater. Water power is derived, from a steady force which does not alter in direction. In contrast, the wind is elusive both in strength and direction, which makes the harnessing of it much more difficult. Moreover, the waterwheel is protected against excessive play of force by opening or closing the sluices, but who shall temper the wind to the windmill? It must stand four-square to whatever wind may blow.

In size, the very early windmill was a miniature of its successor of later days. It was erected upon a tripod set in the ground to raise the structure sufficiently for the sails to swing clear of the ground, but since these had a spread of- only some twelve to fifteen feet, the overall height of the structure was low. At first the miller worked from outside the mill, since there was only room inside for the millstones, which were, at that time, little more than a large-size quern. The canvas on the sails could readily be set by the miller whilst standing on the ground.

Being of such small size, the windmill could easily be moved from place to place, and this idea, that the windmill was not a fixture, persisted down the centuries, even when it had become so much larger that a team of up to a hundred oxen were required to move it.

The windmill soon became established as part of the economy of the manor, and, as in numerous other directions, the lord of the manor secured rights and charters from the Crown regarding it. Amongst them was the legal curiosity of "soke right" whereby both villein and freeman of the manor estate were required to bring their corn to the mill for gristing.

Development of the windmill progressed down the centuries. The early quern gave place to the later millstones which, by the Seventeenth Century, had become a very fine piece of precision mechanism, truly remarkable for its accuracy of construction and close control in operation. By this time, also, the process had been developed for dressing the wheatmeal after it came from the millstones, to produce fine wheaten flour.

Since they were to become the sole competitor of the watermill in the preparation of the people's bread, it was to be expected that windmills would grow in size and numbers. The limit of size appears to have been reached with the installation of three pairs of millstones and a flour dresser. Such a windmill, according to whether it was a post mill or a tower mill, would stand from 60 to 80 feet high and carry a spread of sails of from 100 to 120 feet. As to numbers, it has been estimated that at one period there were ten thousand windmills in England. Isolated mills in the countryside would serve the surrounding villages, their radius being limited by the absence of good hard roads so that long haulage could not be undertaken. On the outskirts of the larger towns would be dotted a number of windmills on vantage points of high ground with enough work for each in supplying the needs of the townsfolk.

A further development should be mentioned. Naturally, in the construction of the earlier windmills it was difficult enough to take care of the workings of the mill without worrying too much about external appearances. At the high peak of their profession, however, the millwrights were very jealous of appearances. The waisted tower of the tower mill has a line of beauty of its own, and similarly the roundhouse of the post mill was made to harmonise with the clean upper structure to give a most pleasing outline. But whatever might be the estimation of the onlooker, the miller invariably regarded his own mill with an affection rarely bestowed upon inanimate things. His life was so largely engaged in taking care of her.

Tower Mill, Sutton near Stalham, Norfolk

Smock Mill, Winfarthing, Norfolk (now demolished)

Types of Windmill

❖

IN the 18th and 19th centuries (1700-1900) windmills were to be found dotted all over the English landscape. But not all windmills were alike or served the same purpose. In the Fen country they were part of the drainage system. They came much later than the corn mill but were actually a simpler form of windmill. The sails were made to drive a kind of waterwheel as it were in reverse. The water from the low-lying land was drained into dykes which were a few feet below the level of the river or the main drainage channels by which the water flowed away to the sea. The reversed water wheel lifted the water just those few feet from the dyke to the drainage channel. There were few "works" in this type of windmill and they required little attention. They are similar to the drainage windmills of Holland and were rarely to be found in England outside the comparatively small area of the Fen country around The Wash. Windmills in which wheatmeal and flour were made were very different. They were crammed full of "works" which are very intriguing from an engineering point of view, considering that they were constructed centuries ago.

All windmills are of course required to work no matter from what point of the compass the wind blows. They must therefore be fitted with some means of turning the sails into the wind. In the small modern farm windmill this takes the form of a vane or tail which acts like a simple weathercock. The problem was much more difficult in the old windmill with its heavy sails.

There are two main varieties of the old windmill – the post mill built almost entirely of timber and the tower mill built mostly of brick somewhat like a lighthouse. The tower mill was a comparatively late comer. It was first introduced about the year 1550, but bricks were then hand made and were difficult to obtain, whereas good solid oak was plentiful all over England, and so, although the tower mill was almost luxurious in the space provided for the machinery, the building and development of post mills was continued. The timber structure of the post mill was built around a massive central post which was supported from the ground on a pyramid of oaken beams. This structure housed as many as three pairs of millstones, the flour dresser and all the driving machinery. For the purpose of bringing the sails into the wind, all of this structure was made to turn round the central post carrying the sails with it and, of course, all the machinery inside.

In the tower mill only the wooden cap which surmounted the round brick tower and carried the sails, turned around to bring the sails into the wind. The wooden cap was arranged to turn on a circular iron track seated on the top of the brick tower, and the drive from the sails was taken through bevel gearing down a central vertical shaft to the machinery situated within the tower below.

A further variation of type was the "smock" mill which like the tower mill had a turning cap for carrying the sails but the fixed tower portion was a many sided structure of wood.

Pictures of early post mills show that the supporting pyramid underneath the wooden structure was not closed in. As improvements were made, however, a brick "roundhouse" was built around the pyramid. The roundhouse then became the granary, a most useful place to store the sacks of wheat while it was awaiting its turn to be ground in the mill above.

A number of other improvements which are not to be found in the earlier windmills were introduced at a later period, but the main principles of construction remained unaltered. Amongst the improvements was the introduction of "patent" sails in which moving shutters replaced the canvas covering of the older pattern and did away with the need for reefing the sails in a high wind. Self-winding gear was also a later development. The cap of the earlier tower and smock mills, or the entire timber structure of the post mill had to be turned around by manual effort in order to bring the sails into the wind. The same elementary arrangement was used for all types. Secured to the moveable cap or timber structure was a long wooden pole, the tail pole, reaching almost to ground level. If the wind veered to a different quarter, the miller came down and trimmed the sails to the wind by putting his haunches to the pole and, walking backwards, shoving the mill around. From necessity he developed a strong sixth sense of tore-casting wind and weather.

The self-winding gear kept the sails automatically trimmed to the wind by means of a "fly-wheel" which was in effect another much smaller windmill built on to the cap of the tower mill or the timber structure of the post mill. It responded to the least beam wind and continued to spin until it had racked the sails back again into the wind's eye.

Sprowston Mill

THE windmill at Sprowston near Norwich, Norfolk, was built about the year 1730. For 200 years it withstood wind and weather and was kept at work every day. Only on a calm summer's day when there was not the least puff of wind would her sails be stilled, or on rare occasions when the millwrights were at work repairing them. Many a time, when the March gales blew strong around her, would the miller spend anxious days and nights fearful for her safety, but she outrode them all and eventually it was fire and not wind that destroyed her in 1933.

Windmills were built in Europe as early as the 12th century, so a mere 200 years is only about one quarter of their historical period. Sprowston Mill was very "modern" and represented the highest development to which the millwright's art of construction was carried.

In these days the wheat was ground into wheatmeal by millstones. The wheatmeal was often used to make wholemeal bread just as it came from the millstones, without going through any refining process. Wheaten flour, however, was also made in the windmill by a simple process of dressing or sieving the wheatmeal through a silken cloth, but the refining process was not carried to extremes. Nowadays flour is not considered to be of first quality unless it is superfine in texture and absolutely white. These extreme qualities can only be obtained by dressing and re-dressing the wheatmeal until every vestige of the brown skin of the wheat berry is removed and, with the skin, the germ and other vitamin-bearing parts. As a last refinement the flour is chemically bleached. But if the windmill flour was not quite so white as modem flour it was very wholesome and nourishing. Long after white flour was made in the modern roller mills, many people who had known the windmill flour still preferred it and insisted upon buying it.

Early maps of the City of Norwich show several windmills in the vicinity, but at the beginning of the present century there were only four left standing conspicuously on the outskirts. These, together with one watermill situated within the city, must for long years have supplied the city with wheatmeal and flour.

As the roller mills became established the windmills declined. They could not now supply flour in sufficient quantity, and, further, millstone flour was placed at a disadvantage in comparison with roller mill flour because, at all horticultural shows and exhibitions, the blue riband of merit was awarded on appearance of the bread and confectionery exhibits. Obviously the judges could not be expected to undergo a course of diet, first on millstone bread and then on roller mill bread, to determine which was the more wholesome. Roller mill flour produced the whiter bread and was judged accordingly. And so, one by one, the windmills were dismantled and disappeared from the landscape until only Sprowston Mill on the edge of Mousehold Heath was left. From time to time proposals for its preservation were placed before the City Elders by various individuals interested in the archaeology of the county, but, even though the mill was to be a free gift from the owner, there was always the disturbing question as to whence should come the necessary funds for upkeep. Then, in 1932, the Norfolk Archaeological Society resolved upon formal acceptance of the mill into its keeping. We have seen how that resolve was frustrated by the narrow margin of one day.

Amongst those who were fascinated by the mill and its workings was the late Mr. H. 0. Clarke of Norwich. He was an engineer by profession, and had made a pastime of the study of windmills, both in England and on the Continent, and regarded Sprowston Mill as unique in design and construction. He determined to make a scale model of the mill, and for some years he would pitch camp under canvas on the mill hill for months together in summer, checking details and making sketches. The completed working model is an exact reproduction in minute detail. It has since been acquired as a national treasure and is on permanent exhibition at the Science Museum, South Kensington, London.

Again, over in the National Gallery hangs a picture – "A Mill on Mousehold Heath" – by the Master of the Norwich School of Painting, John Crome. The mill is a small feature in the picture and has been treated with artistic freedom, but there is Sprowston Mill.

The Miller

THIS is the story of the miller. In most cases he owned and worked his mill which remained in the same family for generations. He sometimes had a paid miller to help him and a carter to bring the wheat to the mill and deliver the flour. Since there was no other source of supply, the villages in the neighbourhood of the mill were dependent upon it for their daily bread. Most families baked their own bread and the miller had the important responsibility of keeping them supplied with wheatmeal and flour. He had at all times to make the best use of the wind. The huge sails would turn even in the lightest of breezes and do a little work. He would grind the wheat when the wind blew strong and then if it came a light wind the millstones would be uncoupled and the flour dresser, which required less power to drive it, would be started up.

Wind by night was just as valuable as wind by day and had to be used. The miller was a light sleeper and oftentimes he would turn out at night in any weather which brought with it a good stiff wind. In a gale he would stand by to see that the sails were properly trimmed and all was shipshape.

As a skilled miller he knew how to blend the softer locally grown wheats with the harder and drier imported ones. There was an art, too, in grinding to the best degree of fineness. In order to produce bread of good texture some wheats had to be ground finer than others. His laboratory was the kitchen of the mill house where Mrs. Miller would be called upon to bake special batches of bread to try out the texture of the blend. He always kept a close watch oh his millstones to see they were in good condition – sharp and well dressed. The dressing of a pair of millstones was a subtle art acquired through years of practice and experience.

Then the miller had to be resourceful in looking after his windmill and its gear. The millwrights were called in for big jobs such as refitting the sails, but he did most of his own carpentry and blacksmithing and even darned the silk cloths by which the wheatmeal was dressed to produce white flour.

The surrounding farmers were his close associates and he was almost as much concerned as they themselves were about the yield of wheat at harvest, seeing that his supplies for flour-making came from them. Besides purchasing their wheat he did their gristing work – grinding and crushing their grain for cattle fodder. This was before the days of the small oil engine which most farmers now use. In still earlier days the possession of a wind-mill carried with it what were known as "soke rights," under which it was compulsory for everyone in the surrounding manor to send their corn to the mill for grinding. However, since there were no other means of doing the work, soke rights probably did not take much enforcement.

Altogether the miller's life was a busy and useful one.

The Sails

THE windmill was mostly built on high ground (the mill-hill) to avoid obstructions to the wind – "the mill on the hill" idea. Then the owner often had legal rights over the land nearby which prevented anyone putting up any building higher than a low shed. The wind in the neighbourhood belonged to the miller, as it were, and obstructions to its smooth passage to the mill were not permitted.

By modern standards the mill sails would not be regarded as a very efficient apparatus for converting the energy of the wind into useful work, but it would be wrong to say they were not very effective for the purpose required of them. Their huge spread – some 90 feet in diameter – enabled the lightest breeze to turn them, and in a very stiff wind they would develop about 25 horsepower to drive the mill machinery.

It was not until the beginning of the present century that scientists began to experiment on the resistance offered by the air to the passage through it of fast-moving objects. As a result of their researches, the principle of streamlining is now applied to all such fast-moving objects – aircraft, motor vehicles, locomotives and ships. In a stiff breeze the fast-moving tip of the sails had about the same speed as an express locomotive. One could stand on the mill-hill beneath the sails and hear the powerful swish-swish as each in turn swung down and followed aloft in endless pursuit of its fellows. But the mill sails did not have that sleek streamlined finish of the modern propeller blades. There were necessary projections on both the front and back faces which made them anything but smooth or streamlined.

Each pair of sails was secured to a backbone or "stock" which reached almost from the tip of one sail to the tip of the opposite one. The sketch shows how the stock was fitted through a socket, known as the "poll," cast on the outer end of the hollow main shaft leading into the interior of the mill. On one side the stock had a projection shaped upon it to locate it centrally in the poll. In addition it was secured by oak wedges driven in hard between the stock and the poll.

The stock was reinforced and strengthened along the centre third of its length by means of wooden clamps on either side, which were bolted on outside the poll.

After a high wind the miller would always climb out through the wicket door behind the head wheel on to the sails and take a look at the stock fixings to see that all was secure. Not only does the greatest strain come upon the stock where it fits into the poll but rain water running down the sails would gradually cause the wood to rot. The life of the stocks was round about 20 years. If they were allowed to go too long the whole sail might break off in a high wind. This was a bad disaster. The heavy flying sail would of course smash itself and anything it struck, but, worse than that, the three remaining sails would cause the mill to run backwards and jerk to a sudden stop, causing severe damage to the works.

Whilst everything was in order, the miller got his power at the cheapest possible rate – just nothing at all, but the wind took its toll of payment whenever the sails required refitting. This was an expensive business. After the millwrights had been paid, the wind did not seem quite so cheap for a good while.

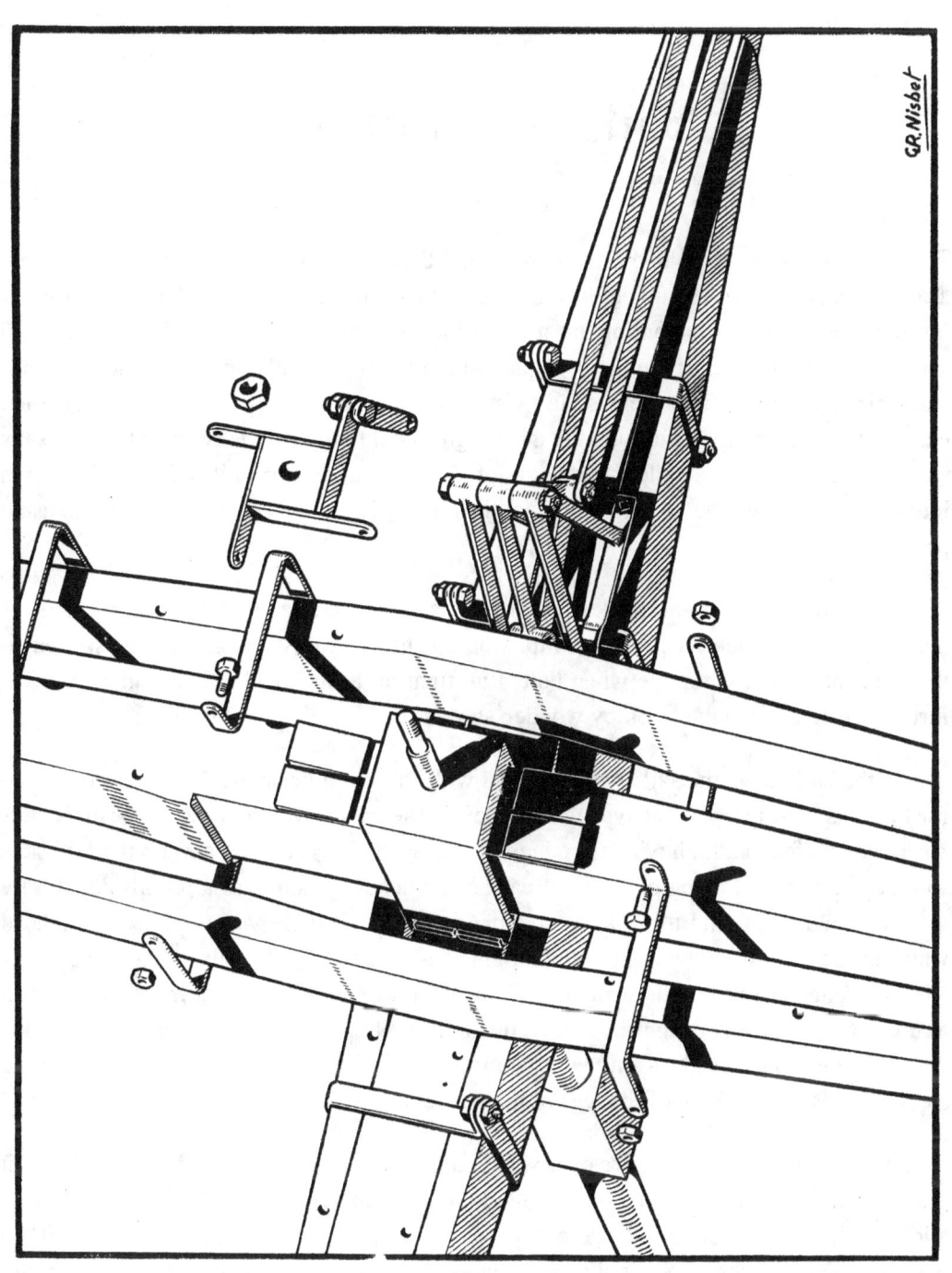

Exploded View of Stocks

Sails and Shutters

IF we examine an aeroplane propeller we find that the blade is twisted more at the root than it is at the tip. The object is to ensure that the fast moving tip and the slower moving root both strike the incoming air at the most favourable angle. So with the windmill sails. The tips of the sails are almost square on to the wind whilst the root is set at a considerable angle to it. But whereas the twist of the aeroplane propeller blade is calculated exactly to a mathematical formula the millwrights could only set the twist of the mill sails by rule of thumb. It was called the "angle of weather," and every millwright had his own idea of what it should be. A figure of 7" at the tip and 20° at the root would be about usual practice.

The light wooden framework of each sail was built on to a central member known as the "whip." The back face of the whip, where it fitted against the stock, was shaped so that it set at the correct angle when bolted up tight on to the stock. The framework was further stayed back to the stock by wooden stays.

In the early days the sails were covered with sail-cloth or canvas similar to that used for the sails of ships. The canvas was tied on to the framework and just as a ship's sails have to be reefed in a high wind to reduce sail so the miller had to swarm up the mill sails and take in a reef all along one side of each sail whenever there was a promise of stormy weather. In a high wind this was a most dangerous job. There was a brake on the head wheel inside the mill, but gusts of wind might cause the sails to drag the brake. The miller would be outside on the sails struggling in the wind to reef the canvas. If the sails started to creep round he was caught in a most uncomfortable predicament. In a rain storm his discomfort was increased. Rain water would collect in pockets in the canvas, run up the sleeves of his coat and out of the legs of his trousers.

If he misjudged what the weather was going to do and delayed reefing the sails until the wind blew a gale, the situation was likely to become desperate. He had the choice of allowing the sails to get up a dangerous speed or trying to curb the speed by means of the brake on the head wheel. Since this was a wooden band contracting on to the wooden rim of the head wheel, there was an acute danger of fire starting from the heat generated there. Too much wind was a great trial.

At a later date "patent" sails were introduced. These were an untold blessing to the miller. Instead of the framework being covered with canvas, it was fitted with pivoted "shutters" very much like a Venetian blind. The shutters were closed or opened, as we shall see later, from inside the mill. To stop the mill, the shutters were opened wide. Then the wind would blow straight through the sails and there would be little tendency for them to turn. When the miller wanted to start the mill he closed the shutters so that the sails presented a flat surface to the wind and the greatest turning effort was obtained. He could regulate the closing gear to hold the shutters tightly shut or only moderately tight. Strong gusts in the wind would force the shutters partly open and "spill the wind" so that the sails would not be whirled around at a dangerous speed. The shutters acted, in effect, as a governor which, in a strong or gusty wind, controlled the top speed of the mill.

Gone were the days of swarming over the sails to reef the canvas. Now the wind was much better harnessed and the miller's work and anxiety for the safety of his mill were very much relieved.

How the Shutters Operated

WE have seen how the old canvas-covered sails gave place to the more modern "patent" sails fitted with shutters on the principle of a Venetian blind.

Each shutter was about 12 inches wide. One long and one short shutter made up the 8 ft. width of the sail. The shutters were pivoted at one edge so that when closed they made up a continuous flat surface. When opened the wind could blow straight through the sail since only the edge of each shutter was exposed to the wind.

All the shutters on all four sails – a total of about 200 – were closed or opened simultaneously by means of a central rod – the striking rod – which passed from the inside of the mill up the hollow wind shaft and through both sail stocks to the outside. A spidery arrangement of bell-crank levers transferred the motion of the striking rod to other longitudinal rods called "uplongs" which ran down the length of each of the four sails. The uplongs connected all along their length with small individual levers attached to each shutter. When the striking rod was thrust outwards the shutters were opened and vice versa.

Let us look at the operation of the striking rod from the interior of the mill. It passed through the tail end of the main shaft and was coupled to a rack member through a swivel joint. Meshing with the rack was a pinion wheel which was rotated by a chain wheel mounted upon the same shaft. A chain hung in a loop over this chain wheel. When the miller wanted to close the shutters to start the mill he hauled on the chain. He could not, of course, stand there all the time to keep the shutters closed, so he hooked a weight on the chain.

What happens in a gusty wind? The wind is trying to open the shutters, the weight is trying to close them. If the gust is strong enough it will overcome the effort exerted by the weight and force the shutters open slightly.

When the gust has died down the weight will overcome the wind force on the shutters and close them again. There the weight rides, moving slightly up and down all the time, smoothing out the wind and keeping the mill at a steady speed.

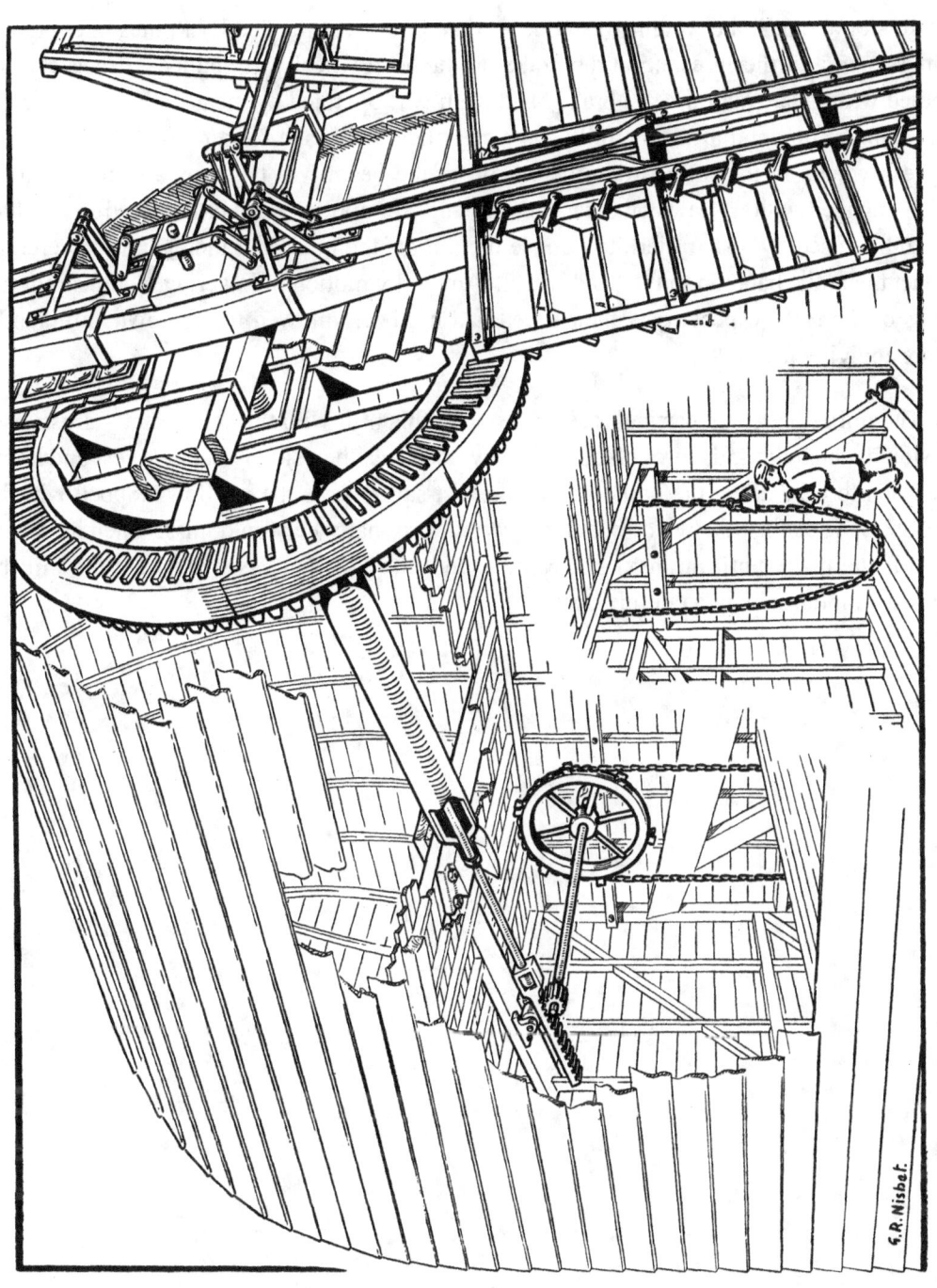

Sails and Shutters

To stop the mill the weight was unhooked and the chain hauled back, thus opening the shutters. As a further precaution the brake on the head wheel was applied. It was always handled with great caution. The brake band itself was operated by means of a heavy timber lever running fore and aft of the mill. Its weight held the brake band tight on to the rim of the head wheel. In the off position the weight of the beam was taken by a swinging hook which engaged with a heavy pin near the free end. In applying the brake the free end had first to be raised by a sharp jerk to bounce the pin off the hook and then gently lowered to contact the brake band on to the head wheel. Any jerky motion in lowering would strain the stocks of the sails, especially if their speed had not been allowed to die down sufficiently before applying the brake.

The miller was very particular about stopping his mill with the sails exactly crosswise. He would watch the sails through one of the windows as they slowly came to rest and knew just when to ease down the brake so that the last sail to pass the window stopped with its edge showing against the window frame. Very occasionally he would miss. Then he would ease off the brake until the next sail came round into position. Never would the mill be stopped with the sails in an untidy position.

Fitting New Sails

THERE is great excitement amongst the miller's youngsters. The millwrights are coming to re-fit the sails. The miller has been a little dubious about one pair of his sails for some time. They have been up some 20 years and the wet has got into them a bit. The time of year is approaching when the weather becomes boisterous and it would be dangerous to let them go another season. It is the "stock" which requires renewal – the timber backbone which passes through the poll on the wind shaft at the head of the mill and to which the sails themselves are bolted.

Already he has been to the timber yard and has selected a fine baulk of pitch-pine timber. The millwrights then gave instructions for it to be rough shaped and delivered to the mill-hill. There it rests on chocks – a stick of timber some 14 in. square in the centre and tapered off to about 7 in. square at each end. It is 75 ft. long and weighs 2 1/2 tons.

The millwrights have planed up the faces of the stock so that the sails will be set at the correct angle when they are mounted upon it. They have also trimmed it so that it will fit snugly into the poll on the wind shaft, and tomorrow they are coming to erect it.

Here is how they go about it. There is considerable preparatory work in removing the sails from the old stock. Guy ropes are first hitched to each end of the two sound sails and hauled upon to bring the old stock to the vertical position. A start is then made with the lowermost sail. The coupling pins connecting the shutters to the uplongs have first to be removed and then the back stays, which connect the sail frame to the stock. Now a tackle is rigged so that the sail may be unbolted from the stock and lowered to the ground. The uppermost sail has then to be brought down for similar dismantling.

One of the guy ropes is hauled upon to bring it oft centre, but since the sails are now out of balance they would swing round with a rush unless gently eased with the brake and steadied with the guy ropes. All of this is steeplejack work and calls for the exercise of care and skill since the sail frames are very flimsy once they are disconnected from the stocks. In due course they are safely lowered and stowed on the mill-hill out of harm's way. After removing the sails the wedges securing the stock in the poll are knocked out and a little rocking motion applied by ropes attached to the lower end soon loosens it and allows it to drop away. So that it does not hit the ground with a thud, a bight around it is taken with ropes secured to the cross sail and it is gently lowered away. When the bottom end reaches the ground, the top end is still sticking a good way through the socket. The bottom end is now levered along the ground and at the same time the guy ropes are hauled until the sails are gradually brought round to about 45°. The top end of the old stock now comes clear of the socket and it is lowered to the ground by blocks and tackle suspended from the cross sail.

The new stock is put in by reversing this process, but seeing that its 2 1/2 tons of weight have to be hoisted, instead of lowered, much more heaving and grunting takes place. When the stock is nearly central in its socket, the sails are hauled round through half a revolution and the new stock drops the last inch or two under its own weight. Projections formed upon it check it by coming in contact with the socket. It is now wedged securely in its socket and is ready for balancing. For this purpose, the sails are hauled to bring the new stock into a horizontal position and any tendency for one end to sink is counteracted by cutting off a piece of the stock from the heavy end or bolting a lump of lead to the light end.

Now comes the re-erection of the sail frames. The stock is again brought to the vertical and the lowermost sail hoisted to it and bolted. The three sails are now out of balance and a considerable heave is required to get the remaining bare stock lowermost. The millwrights make of themselves human ballast by clambering out to the tips of a sail and riding it around.

After the sails have been secured to the stock, there comes the delicate job of re-fitting the back stays to give the correct angle of weather all along the sail from root to tip.

Whilst working on the job, which may take up to a fortnight according to the weather, the millwrights camp and sleep in the roundhouse. They are friendly folk after the manner of sailors, whose work their own resembles somewhat.

To the youngsters it is all very thrilling. They have put their little weight into all the heaving and have held their breath in all the critical situations. They have absorbed all the millwrighting lore and have practised hitches, halt-hitches, bights and bow-lines. Not a move has been missed in the whole of the operations.

The New Stock

The Pyramid Support Structure

O N the mill-hill there is no shelter from the wind be it light, strong or a hurricane. The windmill has both to use the lightest zephyr that blows and to ride out the worst weather the elements can bring to bear upon her. Yet from a picture of an old post mill, it is not at all evident whence she derives the stability to enable her to do so. The upper wooden structure appears to be perched precariously on top of the roundhouse and it looks as if a good puff of wind would topple the whole affair over.

The roundhouse does not take any part in supporting the mill structure. It only serves to enclose the supporting members and acts as a granary or storage space. In earlier post mills, the roundhouse was omitted.

The main supporting structure is in the form of a pyramid constructed of oaken beams. The base of the pyramid rests on four brick piers. Two horizontal beams – the cross trees – are laid with their ends resting on opposite piers. One pair of piers is built up higher than the other pair so that when the beams are-laid upon them they do not require to be mortised at the point of intersection.

Above these beams is erected a pyramid of four stout sloping beams – the quarter bars – the bottom ends of which are chocked into the cross trees. The top ends of the quarter bars converge to meet some 25 ft. above ground level. A massive central post passes up from the intersection of the cross trees through the apex of the pyramid into the interior of the mill above. This post carries the entire weight of the mill structure.

The bottom of the post has two deep mortises cut at right angles to one another to embrace the cross trees at the point where they cross over one another. No weight is carried here however, and in order to ensure that any subsequent settlement does not result in pressure being thrown on to the cross trees a distinct gap, in which the fingers could **be** inserted, is left between the top surface of each cross tree and the bottom of its corresponding mortise in the post.

Unless, then, some other provision were made for taking the weight, the post would slip downwards through the apex of the pyramid until these gaps were closed, and the

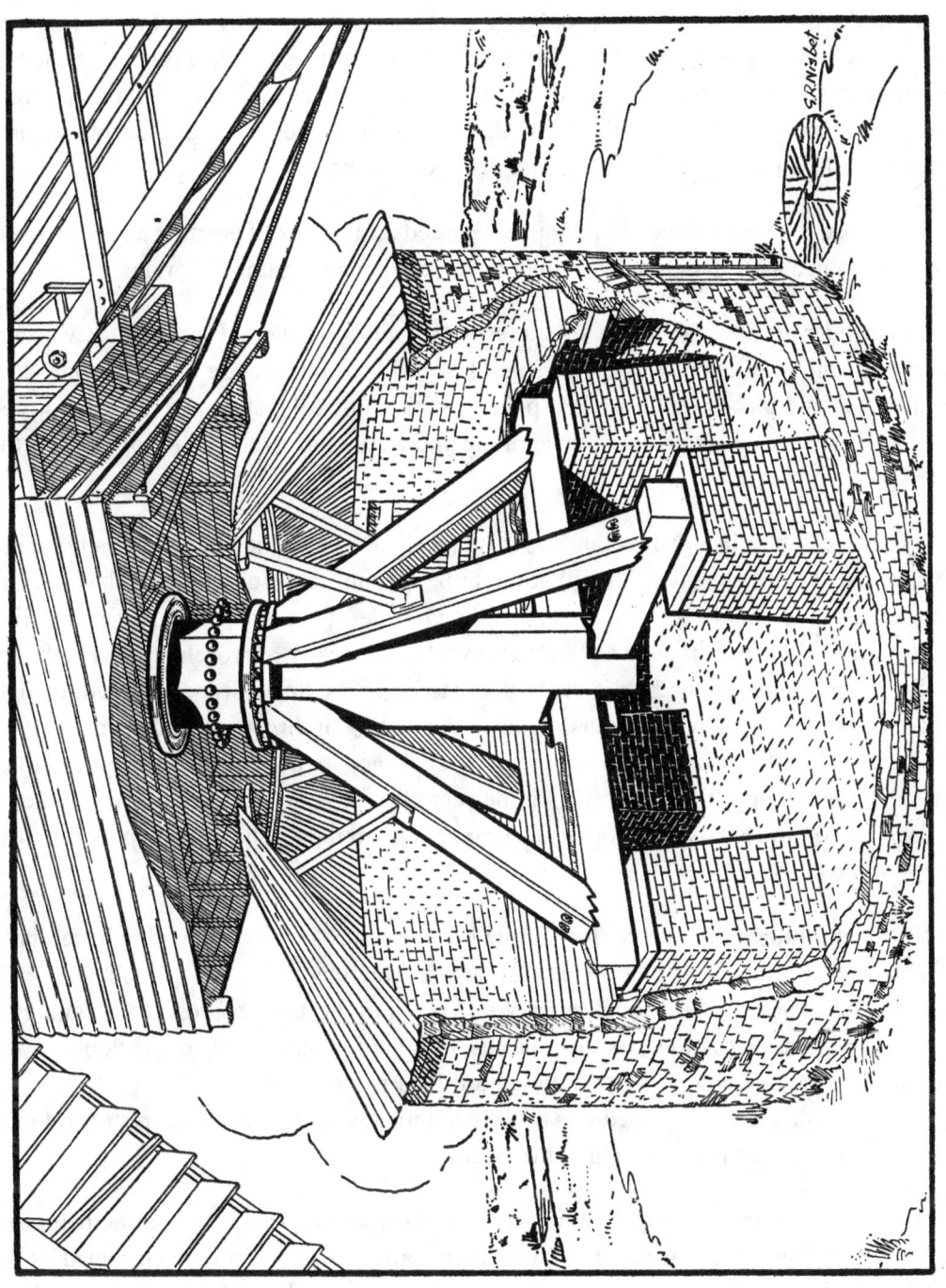

Pyramid Structure

37

bottom of the mortices rested on the cross trees. This is prevented by chocking the quarter bars into the central post at the apex. The weight of the mill resting on a pivot at the top of the central post is therefore transmitted down the post to the apex of the pyramid and from thence down the quarter bars to the brick piers. The cross trees serve to prevent the piers spreading and also to locate the bottom end of the central post.

In a sense the whole structure might be regarded as a means of erecting a post in a vertical position in a more satisfactory manner than merely planting it in the ground.

In considering how the upper structure of the mill is supported by the post we may imagine an empty lidless box to have a small hole drilled in the centre of the bottom. It is then inverted and placed on an upright pointed stick so that the stick pivots in the drilled hole. We then have an elementary model of the upper structure of the mill supported by the central post.

The box can turn in any direction and it is prevented from falling to the ground because its weight is taken by the pivot. It will however be very unstable, or wobbly, sideways, and to correct this we should have to enclose the underside of the box, leaving a neat-fitting hole for the stick to pass though. Now our model more nearly approaches the actual conditions. In the mill itself the box is the bottom storey of the upper wooden structure. The pivot is located in a massive transverse beam in the ceiling of this storey and the neat-fitting hole is formed in the floor of the bottom storey. It is apparent therefore that this storey is suspended (or hanging) from the pivot on the central post. Our model could be completed by placing two further boxes on top of the original one to represent the two upper floors or storeys of the mill.

The "neat-fitting hole" in the floor of the bottom storey is provided with a large diameter ball bearing which encircles the central post and thereby prevents any wobble. Obviously it would not be good enough merely to fix it into the floor boards; it was mounted between two stout joists known as sheer beams, running fore and aft under the floor. It will be clear from the box models that no downward load is taken by this ball bearing and yet, curiously enough, the bearing used was of the end thrust type comprised of upper and lower grooved races with the balls running in the grooves.

This use of the ball bearing is probably the earliest known application of the ball bearing principle. The bearing would nowadays be regarded as very crude, accustomed as we are to the highly finished precision bearing of today. The balls were of cast iron and the grooved races of forged iron not machined in the lathe. Only about one half of the balls carried any load at any one time – the others could be moved by inserting the fingers between

Mr. H. O. Clarke's Model of Sprowston Mill

Marsh (Tower) Mill, Palling, Norfolk (now demolished)

the top and bottom races. Nevertheless, it did its job and so reduced the friction that the earlier mills could be pushed around by hand.

It can now be seen that before the mill can topple over, the central post would have to break off just above the ball bearing or else the whole pyramid would have to be upset from off the brick piers. The post is far too massive and the base of the pyramid is too wide for either of these catastrophes to come about, even in a wind of hurricane force.

The Upper Structure

LET us now look at the upper structure – the mill proper. It is divided into three storeys or floors. The central post extends upwards, from the pyramid structure below, through the bottom floor to the underside of the middle floor. Here is a very staunch oaken beam about three feet square, known as the crown-tree running thwartships of the mill. The crown-tree rests on an iron pivot let into the top face of the central post. The weight of the whole of the upper structure and all its machinery and gear is brought on to the crown-tree. Leaving the crown-tree the first structural members are two curved main beams laid upon it at the outer ends. These are the side girts, each of which carries halt the weight of the mill. At their outer ends they are mortised into the four corner posts which in turn support the three floors. The middle and top floors are, in effect, standing upon the side girts whilst the bottom floor is suspended from them.

On the top floor, under the curved roof, are the big storage bins for the wheat awaiting grinding and for the wheatmeal awaiting dressing into flour. There is very little standing room because the bins take up most of the floor space. The wind shaft which carries the sails comes in at this floor just above floor level. Mounted upon it is the big 13 ft. gear wheel – the head wheel – which is the primary drive wheel for all the machinery in the mill. There is a small wicket door in the front wall of the mill behind this wheel. The miller could crawl through the spokes of the head wheel and out through the wicket door on to the sail stocks and from there clamber on to the sails themselves.

Opposite the head wheel, at the back of the mill, is a look-out window. The mill-hill itself is high ground and the added height of the mill makes of this window an eyrie from which the entire countryside can be surveyed for miles around.

In a strong gusty wind the sway of the mill is very pronounced on the top floor. A stranger would be compelled to hold on to the things to keep his balance but, naturally, the miller has developed sea-legs and is perfectly at home there.

The floor below – the middle or stone floor – is reached by a short flight of steps. All the machinery of the mill is concentrated here – three pairs of millstones, the flour dresser and a grain crusher used chiefly for crushing farmers' grain. The object of placing the millstones on the middle floor is to allow the wheat to run to them from the bins on the floor above and the ground wheatmeal to run away to the bins on the floor below. Similarly with the flour dresser and the grain crusher.

Because of this concentration of machinery, space in the middle floor is very restricted. Open floor space had to be provided on which to set down the upper millstone when it was turned over for the purpose of dressing the surfaces, but there is room for only one pair of millstones to be opened up at any one time.

The next floor down (the bagging floor) is more spacious. The wheatmeal bins are situated here and space is provided near them for swinging the large wooden shovels used for shovelling the meal from the bins into bags. The sail shutters are operated from this floor and also the brake which encircles the head wheel.

The windows are glazed with "bull's-eye" glass. Before it was produced in the form of clear flat sheets, glass tor windows was made by dipping a rod into the molten glass to get a blob on the end of it, and then twirling the rod so that the blob spun out into a circular disc which was afterwards detached from the rod and cut square to the size required for the window pane. This method resulted in a circular lump being left near the centre of each pane. The windows have a very picturesque appearance even if the vision through them is not of the clearest.

When the inspection of factories was brought into force as a legal enactment, the old windmills came within the scope of the act. The inspector's position was not a happy one. He felt compelled to make request for safety precautions which had been non-existent for 200 years or more. He would shudder after clambering up a flight of steps (themselves far too steep and insecure to meet the requirements laid down) to find his head emerging on the floor above within a few inches of a running belt or the whip of the sack hoist chain – hazards which to the miller passed unnoticed and were, in fact, inconsiderable compared with, say, repair work on the sails which the inspector of course never sighted.

As every inch of space was utilised in the running of the mill, if the inspector demanded the usual safety guards and protective measures it would amount to telling the miller he must close up his mill. Honour would be satisfied on both sides by the miller undertaking to replace the loose step where the inspector stumbled, to mend the floppy belt which nearly took off his ear.

If the wind was high on the occasion of the visit, the inspection would, as likely as not, begin and end on the first floor where the little bit of sway would induce the inspector to hesitate before going into the upper regions where it became really pronounced.

Layout – Millstones and Reel Floor

The Wheel the Millwrights Built

JUSTICE would not be done to the old time millwrights without some reference to the construction of the head wheel, or as it was sometimes called, the brake wheel, by reason of the large brake which completely encircled its circumference.

The head wheel was mounted just inside the mill behind the sails on the wind shaft to which the sails themselves were fixed. It transmitted the power developed by the sails to all the machinery in the mill.

Between the sails and the head wheel was the "neck." This was the bearing which supported the cast iron wind shaft and carried the weight of the sails. It took the form of a thick piece of brass shaped to a quarter circle in which the wind shaft rested. The method of lubrication was simple. A large piece of tallow was plonked on the neck brass up against the wind shaft. As the brass warmed up, a little tallow would melt and run in between the shaft and the brass.

It was most important that both the sails and the head wheel should be securely attached to the wind shaft. Since it was difficult in those days to get a satisfactory fixing on a round shaft, other means were adopted. The fixing of the sails was well taken care of by the hollow sockets cast at the outer end of the wind shaft. The sail stocks passed through these sockets and were finally secured by wedges. No slogger could arise here. With the same object in view the round wind shaft was cast with a squared portion where the head wheel was fitted upon it, and this was similarly wedged on to the square.

The head wheel itself was constructed entirely of wood and was truly a masterpiece of the millwright's craft. Every cog in it bore mute testimony to a skill which has now passed into oblivion. Even nowadays the cutting of a bevel wheel of 13 ft. diameter would be a major operation in a machine shop equipped with up-to-date machine tools.

The hollow square at the hub of the head wheel, where it was fitted to the wind shaft, was formed by the two pairs of spokes, arranged like a "noughts and crosses" diagram, and mortised at the intersections. The rim was built up in segments or fellies mortised to the spokes and to one another.

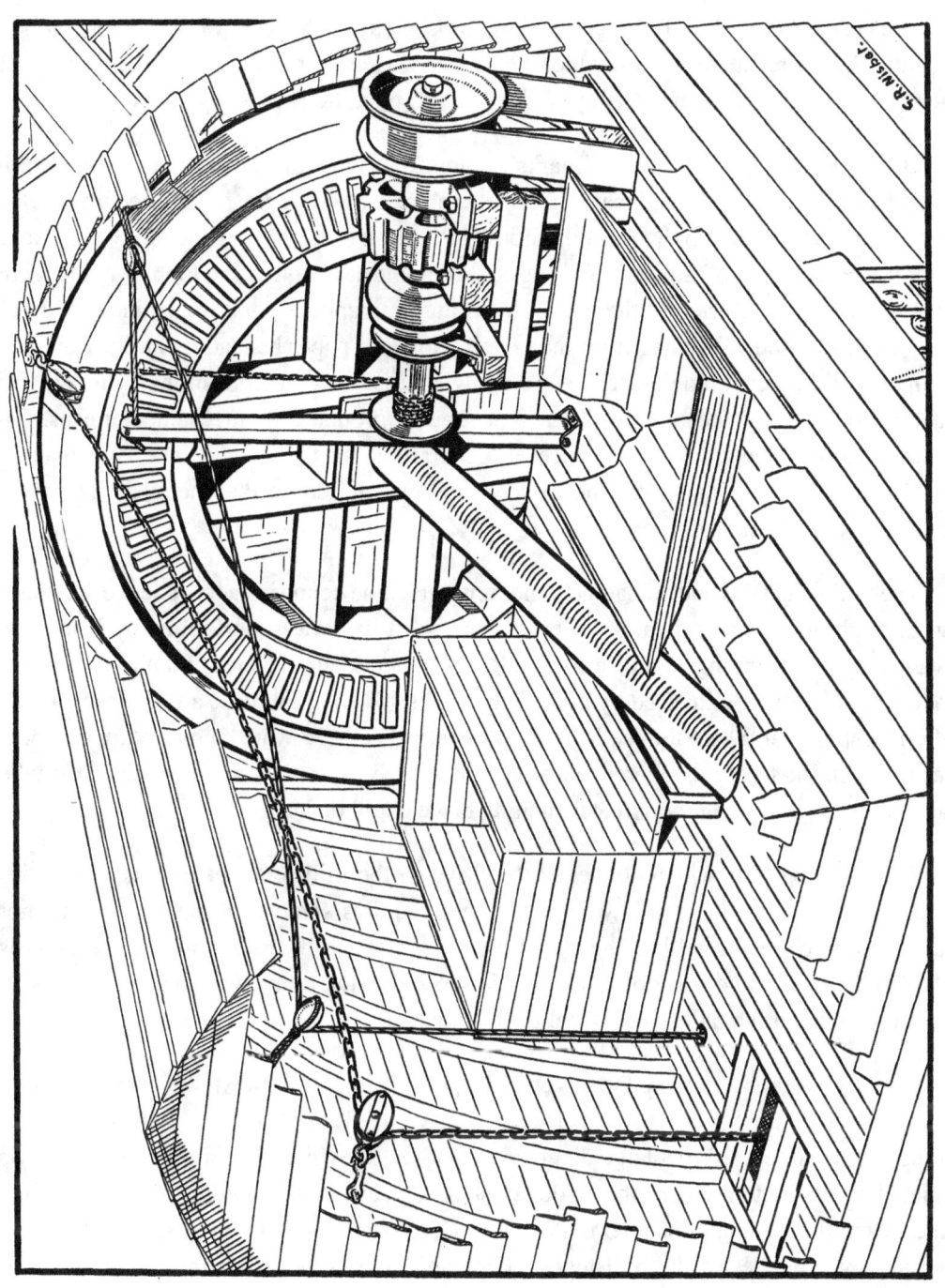

The Head Wheel

The head-wheel was a true cog wheel. That is to say each tooth was a separate wooden cog fitted into a rectangular slot, which was cut right through the rim from front to back. The cogs were secured in the slots by wedges driven in at the back.

On what basis are we to award marks to the millwrights for the accuracy built into the head wheel? In accurately cut gear wheels which are formed by the aid of modern machinery, the teeth are shaped to the form of an involute curve – the curve which may be roughly reproduced by means of a piece of string and a cylindrical object. A loop is made in the string and a pencil inserted in the loop. The string is wrapped round the cylinder and held firmly against it some distance from the pencil. Now the paper beneath is marked by moving the pencil in the path permitted by the string as it is unwrapped from the cylinder. This is a simple curve which can be readily and accurately produced by mechanical means. The process is known in engineering practice as "generating" the tooth form. The same machine which generates the teeth, spaces them exceedingly accurately so that the pitch between all the teeth in the wheel is uniform.

The millwrights of course had no generating machine centuries ago. The work of shaping the teeth and spacing them to an even pitch must have been performed by hand. The advantages of a "hunting tooth" were evidently understood and the difficulty of accurate spacing could only be enhanced by providing it. All this would require a skill of the highest order in a spur gear where the teeth are not tapered. The head wheel, however, was a bevel gear in which the teeth are tapered and yet so accurately were they fitted and cut, that when mated to its meshing gear, no sweeter running drive could be imagined.

The meshing gear which conveyed the drive to the millstones below was a cast iron bevel gear known as the "wallower." This term requires some explanation, as it is more usually taken to mean a rolling about in dirt or mire. It can be used, however, in the sense of "a fading away," and it is obviously in this sense that the term came to be applied, since it is here that the drive fades away from the head wheel.

The wallower was mounted on a vertical shaft which passed down the mill through the middle storey to the bottom storey. The final drive to the millstones was arranged in the ceiling of this storey. Here a large diameter spur gear, known as the Great Spur Wheel was mounted on the bottom end of the wallower shaft. It again was a wooden tooth cog wheel. It meshed with two small diameter cast iron gears, known as the stone-nuts, mounted on the vertical drive spindles of the millstones. Each spindle passed upward through the ceiling and through the centre of the nether stone bedded down on the floor above. The spindles both supported and drove the runner stone of each pair of millstones. Steel keys were used to secure the stone-nuts to the millstone drive spindles – the only place from sails to millstones where this form of fixing was used. A third pair of millstones was driven by bevel gearing taking its drive from bevel teeth inserted in the upper face of the great spur wheel on the wallower shaft.

So far it would appear that the gearing from the sails to the drive spindle was continuous and could not be interrupted. Obviously it had to be so arranged that any pair of millstones could be thrown out of gear when they required dressing or when there was only enough wind to drive one pair. The manner of doing this was simplicity itself:–

There was a detachable segment – the slip cogs – in the rim of the millstone nut. This segment carried five gear teeth and so comprised about one-sixth of the total circumference of the gear wheel.

The gap between the ends of the other five-sixths of the circumference was slightly tapered from top to bottom and the ends also had grooves formed in them. The slip cogs fitted in the grooves and rested securely wedged in the gap under their own weight. To disconnect the millstones the slip cogs were knocked upwards, lifted out of the rim of the gear wheel and laid on a beam nearby. The great spur wheel then ran free in the gap without, of course, turning the millstones. The arrangement might be called elementary but effective.

Besides the wallower one other gear was in continuous mesh with the head-wheel. This was the gear on the cross shaft which at one end drove the chain hoist through a conical friction clutch and at the other end drove the flour dresser by means of a belt pulley.

Viewed from the pinnacle of engineering development reached two centuries later, the mechanism can only arouse admiration for what was achieved with cast iron and wood to produce a geared drive which it would be difficult to surpass for smooth and quiet operation.

The Millstones

ANY idea that a pair of millstones is a rough and ready means of breaking up the wheat and grinding it to a fine flour between roughened stone surfaces would be entirely erroneous. Here indeed is perfection in the stone-mason's art – a perfection of design and construction reached long years ago.

The nether or "bed" stone of a pair of millstones is about 4ft. 6in. diameter and 8in. thick. It is built up in segments of natural rock called millstone grit, cemented together and bound on the outside with iron bands. The bed stone is permanently bedded down in a perfectly horizontal position. The spindle, which supports the weight of the upper stone, and by which it is driven, passes up through the centre of the bed stone. The driving gear wheel is situated below the bed stone up against the ceiling of the floor below.

The upper or "runner" stone is similarly constructed. It is about 14 inches thick and weighs 1 1/2 tons. Right through the centre is a large hole or "eye" (about 12 inches diameter) into which the wheat is fed in a trickle. Across the top of the eye a heavy square section iron bar or bridge is cemented into the stone on either side. The centre portion of this bridge is curved upwards and a recess or dimple is formed on the underside of the curved part to accommodate the pointed end of the driving spindle.

The runner stone hangs, then, somewhat after the manner of a bell, from a suspension point situated just below the centre of the top surface. It has to be in perfect balance when stationary and also when running at speed.

The drive is conveyed from the spindle to the runner stone by means of a coupling element called the "mace." This is a heavy lozenge-shaped piece of iron about 4 inches thick with a square hole through the centre which fits over a corresponding squared portion formed on the drive spindle. The mace, therefore, rotates with the spindle. Jaws which are formed at either end of the lozenge engage with the curved portion of the bridge in the upper stone to complete the drive from the spindle, through the mace, to the bridge.

The function of the millstones is to take in the wheat at the eye of the stones, break up the wheat berry, and then to pass it out at the periphery, ground up into the finest wheat-

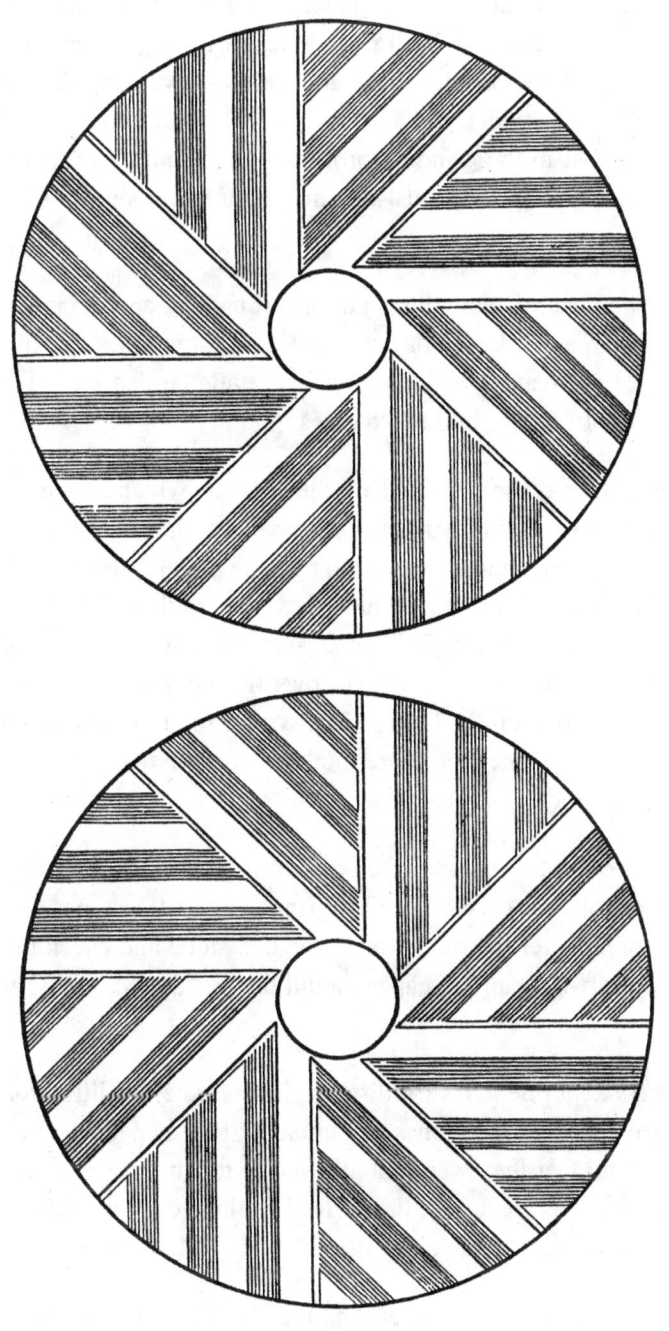

Top: Left-hand Millstone. Bottom: Right-hand Millstone

meal. The broken wheat is made to pass from eye to rim by reason of the pattern cut on the grinding faces. The pattern is formed by a series of deep grooves, nearly tangential to the eye, across the face of the stone. These are the "furrows" and the high parts left between them are the "lands," which themselves have very fine grooves called "drills" cut along their length to assist in the grinding process. The grinding takes place between the lands of the runner and bed stones, whilst the meal is urged along the furrows to the outer circumference.

Millstones may be constructed either right or left handed according as it is intended to drive the runner stone in a clockwise or anti-clockwise direction. The diagram shows the bed stone of a pair of right-handed and a pair of left-handed millstones. It will be observed that the slant of the pattern differs in the two cases.

A good impression may be obtained of the way in which the wholemeal is urged through the stones from eye to rim by tracing each pattern on transparent paper to represent the runner stone. This is how the runner stone appears when turned on its back for the purpose of dressing it. When the tracings are turned over on to their own bed stones the slants will cross over one another. If the right-handed runner is now rotated clockwise, it will be seen how the wheatmeal is urged along the furrows from eye to rim. Similarly if the left-handed runner is rotated counter-clockwise. The wrong rotation would, however, urge the meal towards the eye. The stones therefore may only be driven in the direction for which they have been designed and constructed.

The flat surfaces of the lands are finished off with remarkable accuracy. The "nip" of a good pair of millstones should vary from eye to rim. To test the stones after they had been dressed the miller would lower the runner on to the bed stone and check the nip with paper. At the eye a piece of brown wrapping paper should be just nipped, whilst at the rim a piece of tissue paper.

The miller always kept one pair of millstones in first rate condition for grinding wheat for flour making. He could set the stones to run closer together for the finest grinding, but he had to be very careful that they were not allowed to touch or that the feed of wheat was not allowed to stop. If either occurred, the surface of the stones would be ruined and then would come the long process of re-dressing them.

The wheatmeal which comes out all around the rim has of course to be collected all in one place. For this purpose, the millstones were enclosed in a multi-sided box, known as the vat, a little larger in perimeter than the stones themselves. Small paddles are attached to the rim of the runner stone so that as it rotates they sweep the space between the vat and

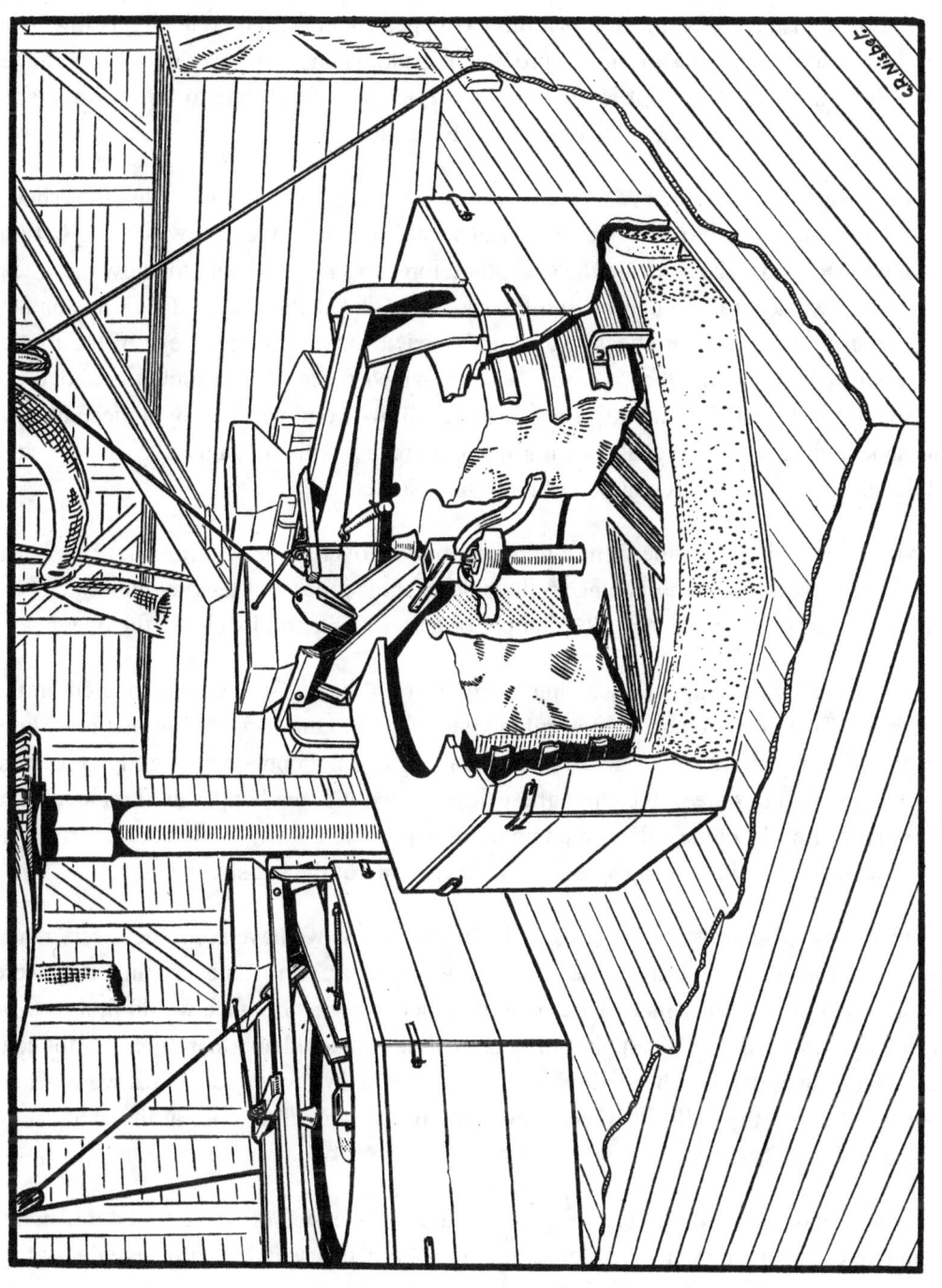

Millstones

the stones and push the wheatmeal into a funnel-shaped spout leading to the bin below. The wheat-meal becomes quite warm in the process of grinding and has a rich pungent perfume. As it collects in the bin below it is necessary to turn it over from time to time with a large wooden shovel to cool it off. It is then ready for bagging.

The wheat was fed to the eye of the runner stone by a sloping trough called the "shoe." It had to be arranged, of course, that the wheat would run when the mill was going but not when it was stopped. The slope on the shoe, therefore, was not sufficient to allow the wheat to run of its own accord. To make it run the shoe was shaken sideways. This was done by means of an iron shaker-arm attached to it and carried forward over the eye of the runner stone. Here it bore against the "damsel" – a four-sided piece of iron mounted above the drive spindle and rotated by it. A springy arm of willow wood was tied by a piece of cord to the shoe and held the iron shaker-arm always in contact with the damsel so that the shoe got four shakings for every revolution of the runner stone.

Wheat was fed to the shoe from a hopper mounted over its backward end. The wheat ran into the shoe through an opening in the side of the hopper, but the amount which was allowed to pass could be adjusted by a damper or gate which partially closed the opening.

In turn the hopper was fed by a chute leading from the big storage bin situated on the floor above. There was no need to control the rate of feed to the hopper. It was merely kept filled by fixing the chute so that its open end was inside the hopper and slightly below the rim. On opening the slide in the chute the wheat gushed into the hopper until its level rose to the open end of the chute, after which no more wheat could flow. Without any further attention the hopper could now draw on the whole capacity of the big storage bin.

The wear on the millstones was very slight because, as we have seen, they were never allowed to rub together. A little of the surface was, however, removed every time they were re-dressed and occasionally nails, wire or some other foreign substance would find its way in with the wheat, and would chip the surface. A windmill could not, therefore, run for centuries without its millstones becoming worn out. One or two worn-out stones were generally to be found on the mill-hill where they made picturesque flagstones at the entrance to the roundhouse.

A warning device was arranged to give the alarm if the feed of wheat to the millstones was in danger of running out. This took the form of a bell suspended above the damsel by means of a cord. The cord was tied to a leather strap located inside the hopper, which fed the wheat to the millstones. Ordinarily the strap was sagged well down into the hopper by the weight of the wheat and this drew the bell away from the damsel. If the wheat ran short, the strap was drawn taut by the weight of the bell which then made contact with the damsel and gave a continuous warning until the hopper was refilled. There would be an interval of three to four minutes from the time the bell began to ring to the time the last of the wheat left the hopper, so the miller just had time to stop the mill or tip more wheat into the wheat bin.

The noise inside the windmill was of the nature of a pleasant soft purr coming from the millstones themselves and the wooden cogs of the driving wheels. Any noise other than this purr immediately put the miller on the alert. The sharp note of the warning bell always caught his ear and sent him scurrying to look after things because of the dire consequences of delay.

Governing the Millstones

THE modern electric motor which drives all manner of machinery runs at a steady speed because the speed of the generator at the power house is itself held steady by means of a governor.

As we have seen, only the top speed of the windmill was controlled by the sail shutters. It ran at anything but a steady speed. Its generator – the wind – could not be fitted with a governor. In a light gusty wind the sails would be swinging with the laziest of motions one minute and running merrily the next.

Without special devices to take care of these variations in speed, it would not have been possible to produce wheatmeal and flour of an even texture and good quality.

Two matters had to be looked after. One, the rate at which the wheat was fed into the stones, and, two, the closeness of the upper millstone to the nether. The miller could set both of them by hand and after he had done so they were maintained in that adjustment automatically. From time to time he would sample the meal as it trickled into the bin from the millstones. He could tell by sleeking off a little meal on a wooden pallette and also by getting the feel of it between his fingers whether to adjust the feed or the closeness of the stones or both.

The automatic adjustment to the feed was very simple. We have seen that the shoe which led the wheat into the eye of the runner stone was continuously shaken sideways by a shaker-arm which bore against the four-sided damsel mounted on the drive spindle. If the mill ran faster the shoe got more shaking and more wheat trickled into the stones. In this way the feed was automatically increased as the speed increased. In similar fashion when the mill ran slower the shoe got less shaking and so less wheat trickled in.

Of course, no more wheat could pass into the stones than was allowed to run into the shoe from the hopper. This was controlled by the gate which partially closed the opening leading from the hopper to the shoe. A cord was taken from this gate to a spool near the meal bin, where the miller could adjust it by hand as he sampled the meal.

The automatic device for controlling the closeness of the stones was more complicated. Let us first see what it had to do. A strong gust of wind makes the sails turn faster. More wheat is fed into the millstones since the shoe is shaken more rapidly. The runner stone is also now rotating faster and is more strongly urging the meal along the furrows from eye to rim. We may now be grinding at the rate of four bushels per hour, whereas previously we were only doing two. In order that the same degree of fineness may be maintained this requires that the stones should now run closer together or, in millering language, that the wheat should be given more of the weight of the runner stone.

All the time that the mill was working in a variable wind the runner stone would be continuously moving closer to or farther away from the bed stone – closer as the speed increased, away as the speed diminished. The movement was very slight, almost microscopic, but when we consider that the wheat had to be ground to powder fineness the need for such movement is more easily seen. Here we have another instance of the delicacy and precision built into the millstone mechanism. Its purpose went far beyond that of "grinding wheat." Without the delicate adjustments provided in the mechanism the texture of the wheatmeal, the flour, and eventually the bread, would vary with the weather conditions at the time the wheat was ground. Hard and soft wheats would produce widely different textures. Unless evenly ground some wheats would be uneconomical for flour-making because of the higher proportion of coarse material to be subsequently dressed out of the wheatmeal in the flour-making process. It must surely be regarded as remarkable that these effects were recognised hundreds of years ago. It is equally remarkable that such clever methods were worked out for controlling to a nicety the movement of a whirling millstone weighing 1 1/2 tons.

We have seen that the runner stone was supported by the drive spindle at the single point where the curved bridge bar of the runner stone rested on the pointed end of the drive spindle. If, then, the spindle is raised or lowered, the closeness of the runner stone to the bed stone will be affected. This has to be accomplished while the stone is being driven hard to its work. The movement was obtained by a compound system of three levers, working in conjunction with a massive flyball governor.

The first lever, known as the bridge-tree, was a staunch oaken beam pivoted to the cross-tree at one end and resting at the other end on a shoe mounted at the centre of the second lever. It supported, at its centre, the pedestal bearing in which the bottom end of the runner stone drive spindle rotated. The second lever, a somewhat lighter beam than the first, was likewise pivoted at one end and was coupled at its other end to the short arm of the third (iron) lever. The long arm of the third lever engaged with the flyball governor.

Strangely, the governor operated in reverse action to that which might have been expected:–

The weight of the runner stone was supported or balanced by the dead weight of the governor flyballs acting through the high leverage of the lever system. When at rest their weight, acting on the long arm of the third lever, held the runner stone away from the bed stone. As they flew wider and wider with increase of speed their downward weight was no longer so effective in supporting the runner stone, which in consequence closed slightly on to the bed stone. The reverse action took place with decrease of speed.

The arrangement, whilst sound and effective in itself, was possessed of another most important advantage. If the governor failed, as it sometimes did when the driving belt broke, the runner stone was immediately lifted away from the bed stone by the weight of the then stationary flyballs.

The possibility of fire was ever-present in the mind of the miller for the sufficient reason that there was no such thing as a small fire in a wooden windmill. Fire was always all-devouring. The fire hazard at the millstones arose from the possibility of the runner stone making contact with the bed stone, as could occur if a nail or other foreign object passed in with the wheat. The runner stone would be tilted thereby out of equilibrium and strike sparks from the bed stone.

It is evident that, from the point of view of fire risk, any failure of the governor threw the millstones, as far as possible, into the position of safety.

Over and above the automatic governing provided by the flyball governor the miller would set the stones by hand to run closer or wider according to whether he wanted a fine or coarse meal. Having set them the governor would hold the grinding to that degree of fineness.

The hand control took the form of an adjustment to the fulcrum point of the third lever. Raising or lowering this point resulted in raising or lowering the short arm of this lever. Since the free end of the second lever was coupled here, the motion was transmitted through the second and first levers to the runner stone.

For the purpose of effecting the adjustment to the fulcrum the millwrights introduced yet a fourth lever, but this is outside and not to be confused with the 3-lever system of the governing mechanism. The fulcrum point was suspended from the short arm of the fourth lever. The long arm was operated by a cord passing over sheave pulleys in the manner of a block and tackle in order to multiply up the movement. The cord was taken to a fixed spool

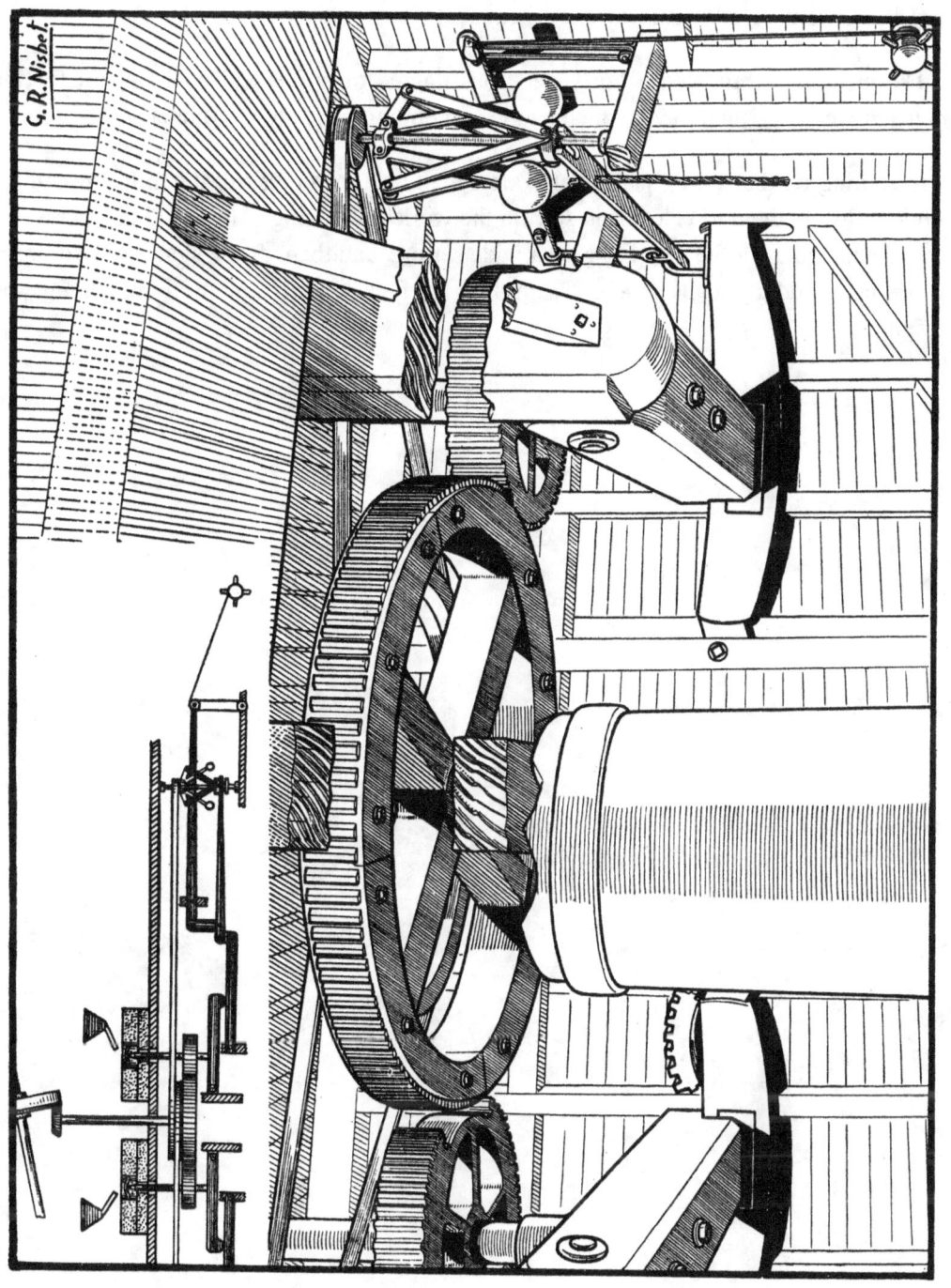

The Millstone Drive and Governor Gear

situated close to the teed spool near the wheatmeal bin.

The mechanism allowed a fairly generous degree of movement at the spool for the minute amount of vertical travel which had to be communicated to the runner stone.

In starting up the stones, particularly after they had been dressed, the miller was most careful not to run them too close and so ruin the work of dressing. He would first screw up the hand adjustment at the spool to give a wide setting, and then, as the stones settled down to work, he would gradually ease off, sampling the meal all the time until everything was just right.

Dressing the Millstones

AMONGST the many routine tasks which fell to the miller's lot was that of dressing the millstones. After grinding some 1,000 bushels of wheat the lands would become worn down until the drills in them disappeared and their surface became glassy smooth. The millstones would then drag badly at their work and would not grind the wheat properly. They were blunt and had to be sharpened or "dressed." This was a job calling for a very great deal of skill and patience.

First of all the heavy upper stone weighing I 1/2 tons had to be lifted away from the nether stone and turned over on its back. This was done by the means of blocks and tackle, but the headroom in the mill was so restricted that it was no easy job. All hands were called in to help. There was always a critical phase in the operation when the stone was just balanced on edge in the vertical position. Its weight then had to be transferred from the lifting slings on one side to lowering slings on the other. It was here that the lack of headroom led to most difficulty.

The tool used for dressing the stones was known as a mill-bill. It was a piece of tempered steel about 1 1/2 inches square and 12 inches long, drawn out at both ends to a fine taper and sharpened to a cold-chisel edge. The miller had a good quantity of mill-bills and all of them had to be sharpened two or three times each time the millstones required dressing. Here was a job for the miller's youngsters – to turn the grindstone while the miller did the sharpening. It was a job at which they would put in a disappearance whenever possible. They didn't like it.

Now the dressing process could begin. The mill-bill was slipped into a wooden holder called a stock. This was no more than a round handle with an enlarged end in which was cut a square hole to receive the mill-bill. When loaded with the mill-bill, it looked somewhat like a long-handled hammer with the mill-bill forming the hammer-head.

The miller would make a few "cushions" out of bags half filled with soft bran and arrange himself comfortably on the millstone, in a half-lying position. Then he would hammer at the lands and re-groove them with the sharp mill-bill. Millstone grit is

61

so hard that to make any impression would take at least a dozen hard hits, all exactly in the same place. A good stone dresser was a skilful and experienced man.

One less experienced would exhibit one of two weaknesses. Either he would not hit hard enough and therefore take an inordinately long time in completing his stone dressing, or he would hit hard but inaccurately and finish up with a badly dressed stone.

Whilst the dressing was proceeding, the surface of the stone would be checked over continuously for high spots and general truth. For this purpose an accurate mahogany straight-edge, known as the "wood-proof," was lightly smeared with moistened red ochre and then struck over the face of the stone. The resultant marking of the stone would guide the miller as to where the heavier dressing was required.

The wood-proof was always handled with much care and respect as being the primary tool necessary for bringing the stone surfaces to that high degree of accuracy required for the production of good wheatmeal. But because it was in the nature of a work gauge and consequently was subject to wear, the miller kept a master gauge which was used only for the periodical checking of the work gauge and not as a work gauge itself. This was the "steel-proof" – a steel bar, one face of which had been finished and scraped to a perfectly flat surface. The steel-proof was kept in its own mahogany case on a particular beam in the mill. The checking of the wood-proof against it was a ceremony carried out twice a year at which the miller always officiated in person. If the wood-proof exhibited high spots they would be carefully rubbed down with pumice stone.

Today it is only those workshops engaged on high precision work which make this distinction between work gauges and master gauges. The age-old usage here of the principle emphasises the precision which was built into a pair of millstones and which the miller had to maintain.

The bed stone would also be tested with the "tracer." This was a wooden arm, which could be mounted upon the exposed millstone drive spindle above the bed stone. A feather from the fowl-yard was attached to the tracer and smeared with red ochre. The tracer was now rotated by allowing the sails to turn slightly. Its marking of the bed stone disclosed any inaccuracies. This was a routine check at every stone-dressing.

Whilst the stones were open they would also be checked for surface damage in the way of pit holes or crumbling. Any bad pits would be filled up with melted flowers of sulphur to restore the surface.

Dressing the Millstones

The furrows also had to be dressed out, but this was an easier job because it was only a matter of chipping away a little of the stone all over the furrow, without having to leave exact grooves in the surface. This is where the miller's youngsters would start their training at stone dressing.

After 4 or 5 days of chipping away – often cold, stiff and cramped, the miller would have worked over the whole surface of both upper and nether stones. They were then ready to be closed up and swing again to their work.

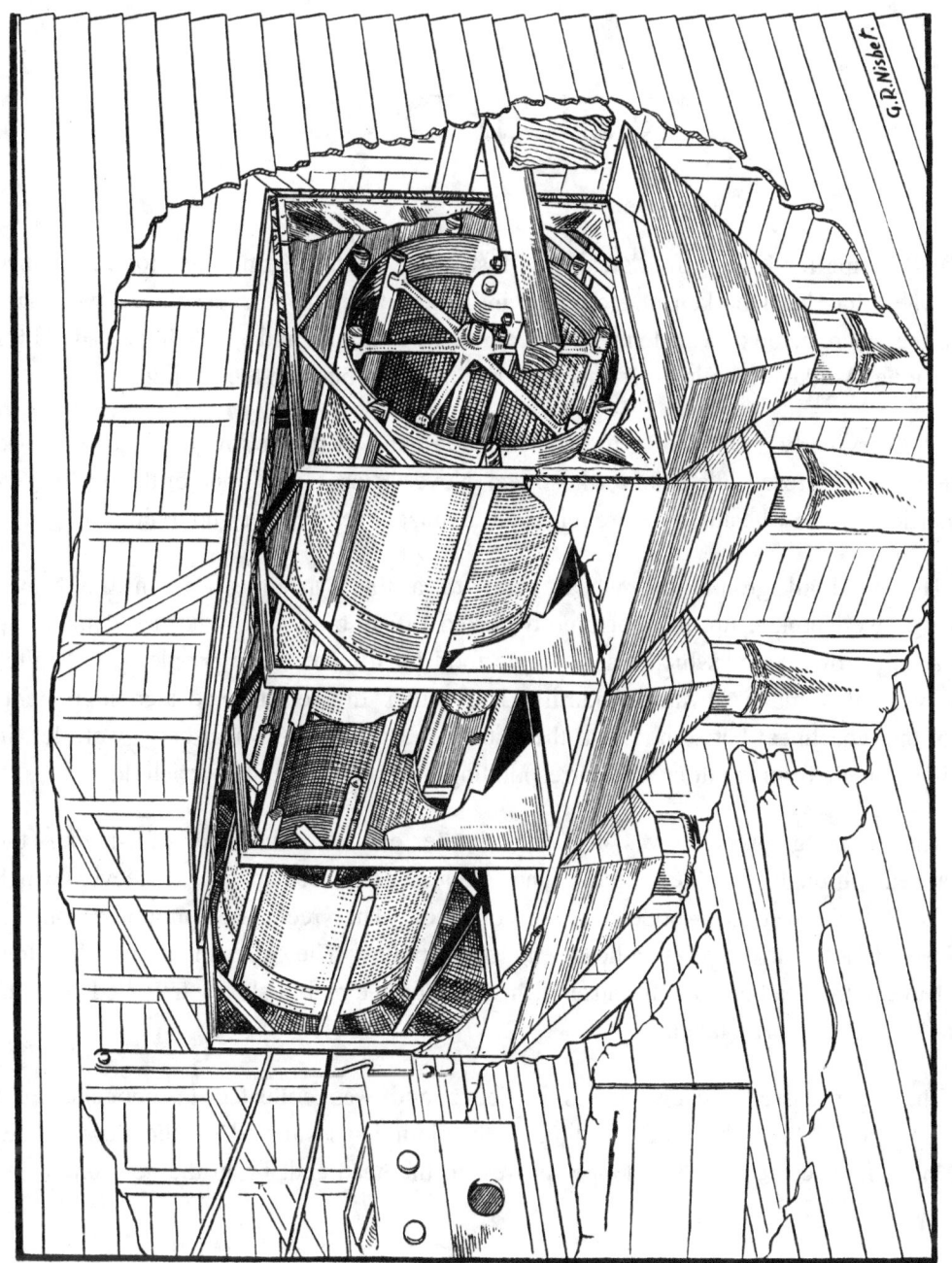

The Flour Reel

The Flour Dresser

THE recipe for King Alfred's cakes is lost in the mists of antiquity. However, since wheat has been broken and crushed into meal from the earliest times, we might guess they were made of wheaten wholemeal mixed with milk and a little honey added for sweetening. A cake tor a King – if only they had not been burnt in the baking.

The very early method of grinding wheat by hand in the hollow of a basin-shaped stone could only result in a coarse wheatmeal. Nevertheless, since none of the health-giving portions of the wheat berry were removed, it must have been very nourishing.

The windmill ground the wheat berry to a fine wheatmeal. Even the brown skin was well broken up and not merely flaked off as bran. The wheatmeal made up into a rich brown bread which was a staple food tor a great many people. But "white" flour was also made in the windmill. The miller dressed the coarser ingredients out of the wheatmeal but some of all the nutritious parts of the wheat berry were left in and bread made from the flour was correspondingly wholesome and of a rich flavour.

The dressing process was a very simple one. It consisted of sieving the wheatmeal through a silken cloth. The cloth was made up as a sleeve, which was stretched over a wooden frame or reel. The reel was mounted on an inclined spindle, driven at the head end by a pulley. The wheatmeal was fed into the sleeve at the head end and because of the slope on the reel, it slowly travelled along to the tail end as the reel rotated.

The flour passed through the silk sleeve and was collected in hoppers, from which it passed through chutes leading to the flour bags situated on the floor below. The finest flour came from the hopper nearest to the head end. From the next one came "second-grade" flour.

By the time the meal in the sleeve had reached the third hopper, the flour content had been dressed out of it. From this hopper came good animal fodder called supers. The coarse bran trickled out from the tail end of the sleeve and was also used as animal fodder.

Now the flour would not pass through the silk sleeve by merely rotating the reel on which it was stretched. Around the outside of the sleeve wooden bars, called beaters, were fixed. As the sleeve rotated, the weight of the meal inside it caused it to belly out and rub against the fixed beaters, so that the flour was actually rubbed through the silk cloth. The beaters became polished to a smooth glossy surface by the constant rubbing.

The drive for the flour dresser was taken from a pulley mounted upon the sack hoist shaft. A belt drive was taken from this pulley to a compound pulley, which was mounted in a swinging frame. Another belt drive was taken from the larger diameter of the compound pulley to the flour reel. To disconnect the drive, all that was necessary was to hoist the free end of the swinging frame, which resulted in the slackening of the first belt drive.

There was little to go wrong with the flour dresser. The silk sleeve was reinforced with leather at the head and tail, where it was lashed to the reel, but occasionally a hole would become worn in it and the miller would have to get to work with his darning needle.

The Fly-Wheel

CLOSE by the mill-hill you almost always found a mast with a weather-cock or wind-vane at its top. The miller was always concerned about the direction of the wind because the safety of his mill depended upon the sails always facing into it. In exceptionally stormy weather the wind might change completely and blow from the opposite direction all in the matter of a few minutes. Then there was danger.

The strongest wind blowing on the face of the sails only forced them more firmly on to the stock to which they were bolted. But if a strong wind got on to the back of the sails, disaster was almost certain. They would now be tail-winded and the safety of the pivoted sail shutters was lost. Instead of the pressure of the wind acting to open the shutters and blow harmlessly through them they were forced to the closed position and the wind blew from behind on the full surface of the sails. They would begin to turn violently in reverse direction and would soon be ripped off the stocks like leaves in autumn.

How were the sails kept into the wind? The mill structure was, of course, so heavy that the simple vane such as is now used on small farm pumping mills would not have been effective. Actually the early windmills had nothing at all to keep the sails into the wind. There was a stout pole – the tail pole – fixed to the mill structure and slanting to the ground. Whenever the wind changed, the miller had to come out and push the mill round by means of this pole. This elementary procedure was used on all the earlier windmills.

Later they were made to be self-winding. In place of the tail pole a wooden structure was built on to the back of the mill (the sails are at the front or head). The wide steps leading into the mill were built into this structure. It was supported on the ground by two track wheels which ran on a circular track of flag stones completely surrounding the base of the mill, concentric with the roundhouse. High up on the structure was mounted a "fly-wheel" (or fan-tail) which was a little windmill in itself. It was connected by bevel gearing to the track wheels so that whenever it turned, it slowly rotated the track wheels and caused them to travel around the circular track.

The main sails of the mill were of course constructed to turn in a head-on wind. The fly-wheel was constructed to turn in a side-on wind. So long as the main sails were facing

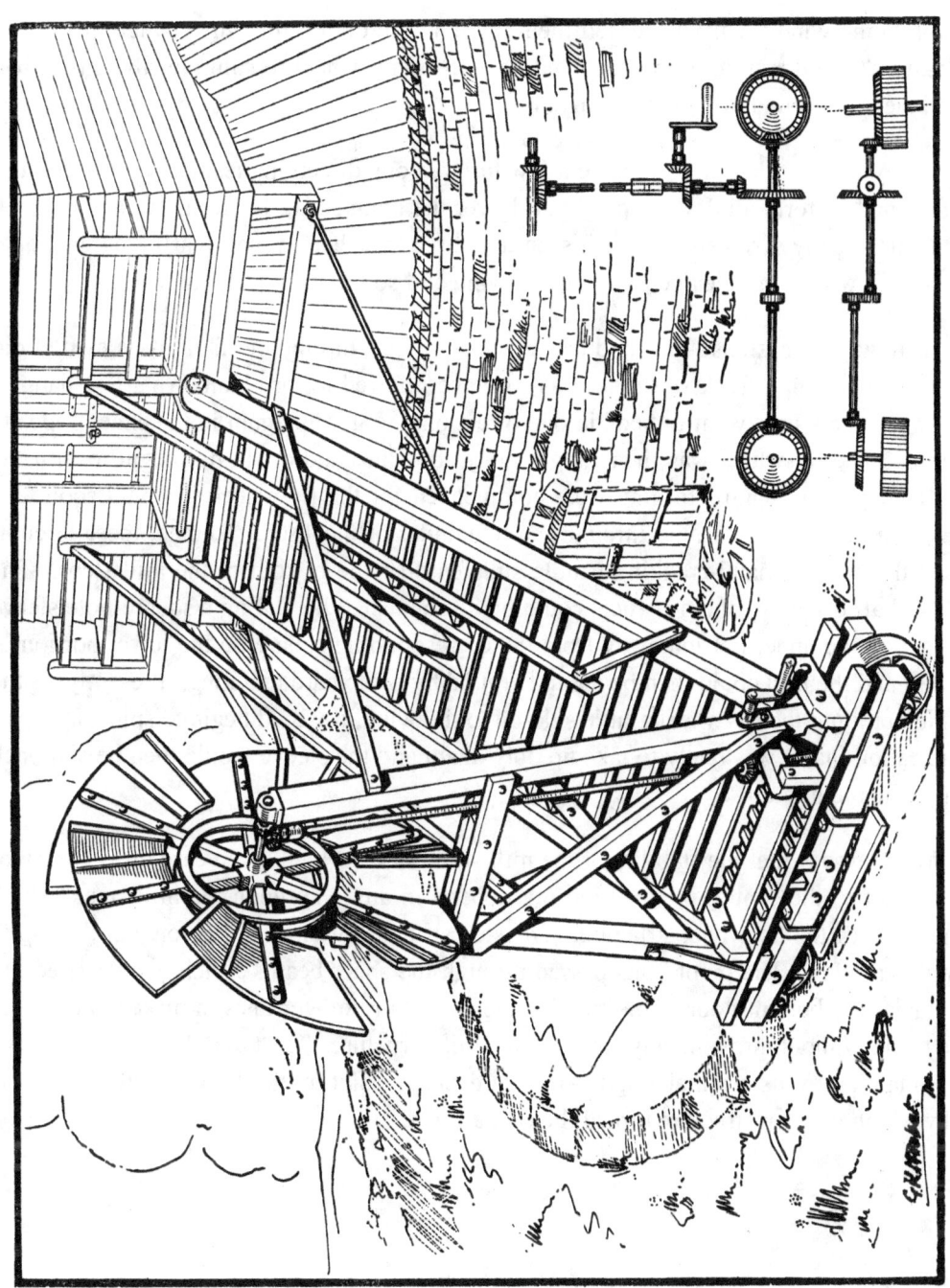

The Fly-Wheel

head-on to the wind the fly-wheel left them there and did not turn. Any change of wind to the least side-on direction, however, would cause the fly-wheel to spin merrily until it had wound the mill around into the wind again.

Mostly the fly-wheel would make a few turns in one direction, hesitate, turn back, turn on again and so forth all day, keeping guard over the main sails. In stormy weather it took the situation really seriously, racing first in this direction, then in that, fearful lest it should not catch up with every change in the strong wind and so fail in its duty as protector.

There were occasions – few and far between – when, through no fault of its own, it did fail; for instance in a freak storm such as we have referred to, or – somewhat less danger-ous – when the wind "went to bed" in one direction and got up again in the exact opposite direction. The fly-wheel was then caught on dead centre and could not begin to turn the mill. If, however, manual turning were commenced for the fly-wheel, it would soon take charge and race the mill round through the remainder of the 180°. To meet these circum-stances there was a hand-cranking handle situated near the bottom of the mill steps. The vertical shaft driven by the fly-wheel could be disconnected by sliding a square-hole sleeve mounted on the upper portion away from the corresponding square on the lower portion of the shaft and the mill could then be moved around by some hard cranking. Presently, as the cranking brought the fly-wheel a little side-on to the wind, it would begin to spin. This was an indication that it was ready to take up duty again and the sleeve would then be slid back again to enable it to do so.

We have seen that the stability of the mill structure was well taken care of by the pyra-mid support to the central post. It did not depend in any way upon the mill-steps, which formed part of the fly-wheel structure. The step structure was in fact pivoted at the top to the mill structure. The pivot bolts passed through the sheer beams which were carried out from under the bottom floor. This enabled the steps to ride over uneven places in the flag-stone track, without throwing any strain on the mill structure. The tail pole – now cut down to less than half of its original length – passed through a slot in the steps to provide a lever-age by which the step structure turned the mill around.

The Sack Hoist

ALL the "works" of the post mill were contained in the upper wooden structure which carried the sails. Underneath was the brick round-house. This contained the pyramid of oaken beams supporting the central post around which the upper structure turned as the wind changed in direction. Not nearly all the room inside the round-house was occupied by the pyramid. Here the wheat was unloaded from the farmers' waggons and stored, waiting to be ground.

There were two floors in the round-house; one at ground level and the other at the level of the top of the brick piers on which the pyramid rested. This was the loading floor from which the farmers' waggons were conveniently loaded.

From the round-house the sacks of wheat had to be taken to the topmost floor of the mill. They were hauled up by the sack hoist which, of course, like everything else in the mill, was operated by wind power. The sack hoist was situated on the top floor and was driven direct from the head wheel. One member of a cone type friction clutch was driven by a bevel gear meshing with the cogs of the head wheel. The other member was secured to a drum around which the hoisting chain was wound. The outer bearing of the drum was mounted in a vertical post which was hinged at its lower end. A handline was attached to the upper end so that when the line was hauled, the two members of the friction clutch were put into engagement and the chain commenced to wind on to the drum. The chain passed over lazy blocks hanging from the roof beams down the mill to the round-house. Trap doors opening upwards were fitted to each floor so that as the sack passed through, the door closed immediately behind it. The handline likewise passed right down the mill. This enabled the miller to send up or lower a sack to any floor, no matter on which floor he then was. He could not, however, deal single-handed with a quantity of sacks.

Let us look on whilst the miller hauls up a waggon load of sacks which have just been delivered to the mill. He is short-handed today because his right-hand man is away carting wheatmeal and flour to his customers. The big wheat bin right at the top of the mill is getting rather empty. Unless more wheat is soon shot in, he will have to stop the mill, otherwise the mill stones will run dry and be very badly blunted. One of the miller's youngsters will have to help. They rather like this job of hooking on the sacks and seeing them disappear aloft into the upper regions of the mill. It is a little bit dangerous and perhaps this gives an added zest to the job.

The two of them go first to the roundhouse where, from amongst the hundreds of sacks stored there, the miller points out the dozen or so that he wants. He then goes to the top-most floor ready to haul on the handline and receive the sacks when they come up to him and to shoot the wheat into the big bin. The only means of communication is now the hand line.

Away down in the roundhouse the youngster is busy with his first sack. It is wedged between some others and is not, by a long way, directly under the trap door in the floor above through which it will have to pass. The end link of the chain is in the form of a ring. He makes a loop by passing some of the chain through the ring and puts the loop over the mouth of the sack where it is bunched together and tied. He sends up his O.K. signal to the top floor – a short tug on the hand line – and soon the chain begins to move upwards. As the slack is taken up, he holds the loop to prevent it slipping off the sack.

Now the pull is firmly taken and the sack begins to drag across the floor until it is plumb under the trapdoor. There is not much room amongst the closely-stacked sacks and he has to be on the alert to dodge the moving sack and avoid being tripped by the chain. Not for worlds would he send up the distress signal – two short tugs. The sack swings into mid-air and passes up through the trapdoor on its journey to the top of the mill. He hopes he has made his loop secure so that it will arrive in safety. He will get no marks if it slips and the sack bursts and sheds the wheat over the floors through which it has passed.

The wind is good today and the mill is running fairly fast so he does not have long to wait before the end of the chain makes its appearance back again through the now open trapdoor. He has got his next sack all ready and presently it follows aloft in the same way. When all have been sent up, he rings off on the hand line with a quick succession of tugs and receives back his "thank-you" in the same manner.

This little help which the youngsters could give was most welcome, and they were al-ways encouraged to take an interest in the workings of the mill. As recompense or sweeten-ing they were permitted many liberties but not, of course, such as would endanger any of the mill gearing and especially the millstones.

There were oftentimes when the winds were light when the mill would be left to itself for almost the whole of the day. At such times no restrictions were placed upon the com-ing and going of the youngsters. These were the occasions when they practised their more

The Sack Hoist

73

daring exploits on the sack hoist. From early days, however, there was a realisation that the economy of the mill-house centred on the windmill itself, and the fact that a blind eye was turned on many unorthodox incidents did not result in liberties being taken.

Besides the attraction of the mill there was the workshop, where some of the tools must surely have dated from the erection of the windmill itself. But even so, the youngsters, having access to them, could scarcely escape becoming budding millwrights.

From yet another angle, the mill had an influence on the youngsters' early outlook. Feeding stuffs were plentiful although officially they had to be purchased from the mill with pocket money. At different periods almost every known domesticated animal must have been bred and nurtured by the youngsters either for its economic value in swelling the pocket money allowance or, quite as frequently, tor the pleasure of cherishing their personal pets.

Painting the Mill

Painting the Mill

WE have seen how the miller was in trust, as it were, of the daily bread of the community living within a few miles' radius of the windmill. In discharge of that trust he had to make the best use of the wind and to keep his mill in good working order. The repair of the sails and the dressing of the millstones have been described as amongst his more important jobs. As may be readily imagined, however, apart from looking after the running of the machinery inside the mill, there was a constant recurrence of repairs to the wooden structure, sails and self-winding gear, seeing that they were exposed to all weathers, winter and summer alike.

One of the bigger jobs was that of painting, which was generally carried out by the miller and his men with not inconsiderable help from the youngsters. The mill was not only a conspicuous feature of the landscape; it was in the nature of a symbol representing the security of the supply of bread to the neighbourhood. The miller was zealous that it should always have a tidy and even smart appearance. The sails were never askew when stopped at night, but always correctly set at 45°.

To maintain a smart outside appearance the mill was painted all over every second year. Combined with his other activities, the miller had to be a steeplejack of no mean consequence. There was plenty of foothold on the sails, although no chances could be taken when, as sometimes happened, it was more convenient to paint a sail in its topmost position.

The painting of the sails could be done piecemeal because no special rigs were required; the shutters provided a ready-made ladder. For painting the sides of the mill and the front and back, a plank had to be rigged for whoever was painting to sit upon. Ropes were slung over the top of the mill and secured to the plank as it rested on the ground. The ropes were then hauled and secured to an anchorage, which was usually the mill waggon, on the other side.

The painter reached the plank by climbing up the sails to the rounded roof of the mill and then sliding down the ropes.

After a concerted effort by all hands for a couple of weeks, the whole of the wooden structure would be painted. Now remained the roundhouse, which was given a coat of hot tar applied with long-handled brushes to reach all over the sloping roof.

The inside of the mill was likewise given a thorough spring cleaning. As a place in which food is prepared, it was always kept in clean condition, and this was one of the occasions when special attention was given to preserving that condition.

There she stands, gleaming white against the sky, ready again to brave the elements which she harnesses and makes to serve her. A goodly servant and a hard taskmaster in one.

GLOSSARY

Angle of Weather: The angle the sails make with the plane of rotation. It varies along the sail from tip to root.

Beaters: The fixed wooden bars surrounding the Flour Reel against which the Silk rubs as it rotates.

Bed Stone: The lower stone of a pair of millstones.

Brake: The heavy built-up wooden band which encircles the greater part of the circumference of the Head Wheel.

Brake Beam: The heavy beam hauled upon by a block and rope to operate the Brake.

Brake Wheel: The Head Wheel.

Bran: The skin of the Wheat Berry, separated from the Wheatmeal in the process of dressing - it to make flour.

Bridge: The square iron bar cemented in across the eye of the upper millstone. It takes both the weight of the stone and the drive from the spindle.

Bridge Beam: The beam on which the toe or lower end of the millstone drive spindle rests in its pedestal bearing.

Bull's-Eye: The old-fashioned glass used for window panes. Mostly has a blob at the centre where it was attached to the rod used in twirling the molten mass to form a flat sheet.

Cap: The wooden structure resting on the top of the brickwork of a Tower Mill, or the equivalent wooden structure of a Smock Mill. Carries the sails and fly-wheel and turns to keep the sails in the wind's eye.

Chain, the: The hoisting chain operated by the Sack Hoist.

Clamps: The timber reinforcements bolted on to each side of the sail stocks.

Cogs: The teeth of the Head Wheel or Great Spur Gear. (See also Pegs.)

Coomb: Four bushels of any kind of grain. Not used as a measure of wheatmeal, flour, or other ground product.

Cross Trees: The two horizontal beams in the Pyramid structure. They rest on the Piers and support the Quarter Bars.

Crown Tree: The massive beam resting on the Pivot at the top of the Post. Carries the entire weight of the mill and everything in it.

Damper: The iron plate closing the aperture leading from the Hopper to the Shoe.

Damsel: The four-sided iron member fitted to the drive spindle at the eye of the millstone. It shakes the Shoe which conveys the wheat from the Hopper to the millstones.

Drills: The fine grooves cut in the Lands of the millstones.

Eye: The circular hole at the centre of the upper millstone.

Fan-Tail: The Fly-Wheel.

Flag Stones: The circular track surrounding the Roundhouse on which the Track Wheels of the Steps structure run.

Flour: Refers to millstone flour made by dressing the wheatmeal in the Flour Dresser.

Fly, Fly-Wheel: The circular wind wheel mounted on the cap of a Tower or Smock Mill or the steps structure of a Post Mill. Automatically keeps the sails in the wind's eye.

Furrows: The deep grooves in the face of the millstone along which the wheatmeal is urged from eye to rim.

Gate: The Damper

Germs: The bud of the Wheat Berry from which the root and stem of the plant grow.

Governor: The fly-ball governor which regulates the closeness of the runner stone to the bed stone.

Great Spur Gear: The cog wheel which drives the Stone Nuts in the ceiling of the bottom floor.

Head Wheel: The cog wheel on the Wind Shaft. The first wheel in the mill. Transmits the power from the sails to all the machinery.

Hopper: The container, shaped like an inverted pyramid, which receives wheat from the storage bins and feeds it to the Shoe.

Horse: The wooden framework supporting the Hopper and Shoe.

Lands: The higher ridges between the Furrows of the millstones. The Lands have fine grooves cut in them in the dressing process and are then said to be sharp.

Mace: The lozenge-shaped piece of iron which fits on to the millstone drive spindle and transmits the drive to the Bridge of the runner stone.

Mill: The entire mill structure. Also used to denote the upper timber structure of a Post mill. (One could be in the roundhouse and say "I am going up into the mill.")

Mill-bill: The steel tool used for dressing the millstones.

Mill-hill: The high ground on which the windmill stood.

Millwrights: The designers and constructors of the windmill. Also the experts who carried out major repairs such as re-fitting the sails.

Neck: The turned and polished portion of the windshaft between the Poll and the Head Wheel.

Neck Boss: The brass bearing in which the Neck runs.

Nether Stone: The lower (fixed) stone of a pair of millstones.

Nip: Defines how closely the upper stone will rest upon the nether after dressing.

Pallet: A small flat piece of wood kept near the meal bin for sleeking off a sample of meal for the purpose of examining its texture.

Patent Sails: Sails fitted with automatic shutters. They superseded sails fitted with sailcloth which had to be reefed by hand in a high wind.

Pegs: The teeth of the Head Wheel and Great Spur Gear. (See also Cogs.)

Piers: The brickwork upon which rests the Cross Trees.

Pivot: The iron bearing which locates the Crown Tree on the Post.

Poll: The square box-like sockets on the wind shaft through which the sail stocks pass.

Post: The central vertical timber upon which the weight of the mill is taken and around which it turns with change of wind.

Pyramid: The timber structure supporting the Post.

Quarter Bars: The inclined beams of the Pyramid.

Rack: The toothed bar, connected to the Striking Rod by a swivelling joint, at the toe of the Wind Shaft.

Reel: The rotating frame of the Flour Dresser, over which the Silk is drawn.

Roundhouse: The circular brick building beneath the timber structure of a post mill. Encloses the Pyramid.

Runner Stone: The upper (rotating) stone of a pair of millstones.

Sack of Flour: 20 stone.

Sail Frames: The framework which is bolted to the Stock and in which the shutters are fitted.

Sheer Beams: The two joists, one on either side of the Post, on which the floor of the bottom storey is laid. They support the Ball Race and the Tail Pole.

Shoe: The wooden trough leading the wheat from the Hopper to the eye of the millstones.

Shutters: The "Venetian blinds" fitted to Patent Sail frames. (See also Vanes.)

Side Girts: The two beams resting on either end of the Crown Tree. They support the corner posts of the mill.

Silk, the: The silken sleeve which was drawn over the Reel. The flour was beaten through the Silk as it rubbed over the Beaters.

Slip Cogs: That part of the rim of the Stone Nuts which is removable for the purpose of throwing the millstones out of gear.

Smock Mill: Somewhat like a Tower Mill. Surmounted by a Cap, but built of timber.

Soke Rights: Legal rights enjoyed (!) by the old-time miller whereby the farmers of the manor were compelled to send their corn to the mill for gristing.

Spider: The bellcrank levers on the outside of the sails near the Poll. They transmit the thrust of the Striking Rod to the Uplongs to operate the Shutters.

Spindle: The drive spindle of the millstones.

Steel-proof: The master straight edge used for checking the Wood-proof.

Steps: The wide stairway leading from the ground to the bottom storey of the mill.

Stock: The main timber which passes through the Poll and to which the Whip is bolted. Also, the wooden handle in which the Mill-bill is held whilst dressing the millstones.

Stone Nuts: The cast iron spur gear mounted on the millstone drive spindle.

Striking Rod: The rod passing up the centre of the Wind Shaft to operate the Shutters.

Supers: A product obtained from the dressing of Wheatmeal. Comes between flour and bran. Excellent fodder for pigs, poultry and cattle.

Sweeper: The small paddle attached to the rim of the Runner Stone. It swept the meal from the Vat into the chute leading to the meal bin.

Sweeps: The Sails.

Tail Pole: The pole secured to the Sheer Beams, which comes out through the slot in the steps structure.

Tracer: The arm temporarily fitted on to the millstone drive spindle for checking the truth of the bed stone when its surface is being dressed.

Track Wheels: The ground wheels of the Step structure.

Uplongs: The rods running down the length of the sails and connecting the Spider to all the Shutters.

Vanes: The Shutters. Also the segments of the Fly-wheel.

Vat: The wooden casing surrounding the millstones.

Wallower: The cast iron bevel gear meshing with the Head Wheel.

Wheat Berry: The grain of wheat.

Wheatmeal: Finely ground wheat. Used without further processing for making wholemeal bread. When processed in the Flour Dresser yields Millstone Flour, Supers and Bran.

Whip: The backbone upon which the Sail Frames are built. It bolts on to the Stock.

Wind Shaft: The main shaft upon which are mounted the Sails and the Head Wheel.

Wood-proof: The mahogany straight-edge used for checking the surfaces of the millstones when they are being dressed.

www.ingramcontent.com/pod-product-compliance
Lightning Source LLC
Chambersburg PA
CBHW081744220526
45468CB00008B/2233